Rethinking Technologies

Rethinking Technologies

Edited by
Verena Andermatt Conley
on behalf of the
Miami Theory Collective

University of Minnesota Press
Minneapolis
London

"The Age of Paranoia," by Teresa Brennan first appeared in *Paragraph* 14, i (March 1991), 20-45. It is hereby reprinted with permission.

Published by the University of Minnesota Press
2037 University Avenue Southeast, Minneapolis, MN 55455-3092
Printed in the United States of America on acid-free paper

Library of Congress Cataloging-in-Publication Data

Rethinking technologies / edited by Verena Andermatt Conley, on behalf
of the Miami Theory Collective.
p. cm.
Includes bibliographical references and index.
ISBN 0-8166-2214-0
ISBN 0-8166-2215-9 (pbk.)
1. Technology—Philosophy. 2. Technology—Social aspects.
I. Conley, Verena Andermatt. II. Miami Theory Collective (Oxford, Ohio)
T14.R44 1993
601—dc20 93-22415
CIP

IN MEMORIAM

Félix Guattari

(1930-92)

Félix Guattari (*left*) and Patrick Clancy during the Technology Colloquium, Miami University, October 1990

Contents

Preface ix

Acknowledgments xv

Part I. Questioning Technologies

1. The Third Interval: A Critical Transition
Paul Virilio 3
2. Machinic Heterogenesis
Félix Guattari 13
3. War, Law, Sovereignty—*Techné*
Jean-Luc Nancy 28
4. Our Narcotic Modernity
Avital Ronell 59

Part II. Technology and the Environment

5. Eco-Subjects
Verena Andermatt Conley 77
6. Age of Paranoia
Teresa Brennan 92
7. Heidegger and the Rhetoric of Submission: Technology and Passivity
Ingrid Scheibler 115

Part III. Technology and the Arts

8. Technical Performance: Postmodernism, Angst, or Agony of Modernism?
   Françoise Gaillard 143
9. The Technology of Death and Its Limits: The Problem of the Simulation Model
   Scott Durham 156

Part IV. Technology and Cyberspace

10. The Seductions of Cyberspace
    N. Katherine Hayles 173
11. The Leap and the Lapse: Hacking a Private Site in Cyberspace
    Alberto Moreiras 191
12. Telefigures and Cyberspace
    Patrick Clancy 207

Works Cited 233

Contributors 241

Index 245

Preface

Verena Andermatt Conley

The essays collected in this volume intend to do just that: to rethink technologies in an advanced technological age. As electronic communication and accelerated modes of transportation shrink our planet more and more, technologies are often assumed to be the science of either salvation or human damnation. On the one hand, postmodern celebrations of contemporary technology and related cultural sensibilities as the most varied, mixed, and "advanced" assert that they are so beneficial they even help women and other cultural minorities gain higher status. They accomplish what humanistic discourses could never do. On the other hand, elegies on the death of nature and the dangers of automation and dehumanization counter the expression of praise.

The essays in this volume try to avoid one or the other extreme. They neither blindly espouse consumerism or snob appeal of "high tech" nor fall into a nostalgic lament forever bemoaning a lost paradise. They look critically at the rapport that is changing in new ways between technology and the humanities. This rapport is complex. Far from simply lamenting the loss of humanness through technology, the essays in this volume attempt to rethink the subject in the wake of a becoming technological of the world. This rethinking takes place in philosophy, psychoanalysis, and the arts. Testing the limits between the humanities and sciences, it touches upon many contemporary issues and political urgencies, such as pollution, war, drugs, consumption, massification, the media, and virtual reality. Gathered around intellectual affinities of the authors, as well as clusters of specific interest, the essays intersect with and refer back to one another. The four parts of this volume, "Questioning Technologies," "Technology and the Environment,'' ''Technology and the Arts,'' and ''Technology and Cyberspace,'' are intended primarily as markers for the reader. If individual articles choose to foreground one or the other of these categories, they are nevertheless all interrelated. It is the interplay among the various essays put side by side that complicates the chosen topic. No linear progression is intended. A slight dyssymmetrical framing effect can be derived from the first and the last pieces, that is, between Paul Virilio's *état présent* of a planet polluted by technologies and Patrick Clancy's transcription of a performance in text and images. In this last piece, a technological creature, a reincarnation of Boris Karloff, shuffles through the world that Virilio has leveled, in order to ratify it, but also to provide certain openings onto new spaces of creation and analysis.

In the research in the humanities and social sciences over the past two decades, technology has been initially defined as *instrumental*. It pertains to human creations that alter or manage what is construed to be the natural environment. Seen in another light, it is thought, in the wake of Martin Heidegger's speculations on science and poetry, to be a *means of execution*, ranging from an ''application'' to a generalized technics, that is, to something that is ''always already'' there and that could be said to circulate ineffably.[1] Following Heidegger, we are compelled to ask not only what our positions are in relation to technology, but also, how and where do we locate the latter? Everything is somehow technical, and technics appear to be a defining trait of all investigation and knowledge. Technics must be distinguished from *instrumental* technology that amounts to one of its particular uses and that consequently has been tied to specific ideologies.

The surging of instrumental technology can be dated historically. It is linked to the second scientific revolution that took place at the end of the Renaissance, and coincides, if Heidegger's vocabulary is still appropriate, with a *techné* or means that flattens a three-dimensional world into a two-dimensional diagram or map, institutes a separation between subject and object, and inaugurates the quest of the rational, self-possessed subject that soon expands and colonizes. In brief, it is one with the ''Western project'' or *projet occidental* that—though initially linked to a geopolitics—is now ubiquitous and has become synonymous with ideology.[2] The stance leads the West to develop a *techné* in the sense of an instrumentality that takes over, arrests, or enframes what it desires to manipulate or contain. Because it has been meant to comprise not merely a sum of beliefs but also the ''imaginary relations that human subjects keep with real modes of production,'' the ideology of technology can be felt in the ways that scientific discourses tell the body how it should act, feel, and live the life that destiny allots to it. The body is subject to the effects of a rhetoric of technical reason.

Yet this body is also transformed by its encounters with technologies. At stake is perhaps no longer, as Heidegger once saw it, a dehumanization, but a transformation of the body and of subjectivities. And it may be possible to rethink technologies in terms other than enframing. Indeed, ecological imbalances have shown that technology does not just master nature. And in its most advanced stages, at the cosmic frontiers, technology reveals the very uncertainties of human thought. Thus, technology not only alters human subjectivities, but, paradoxically, decenters humans' position in the world.

Technologies are responsible for many of the ways we lead our lives. They define our public as well as our most intimate and private spaces. In our daily lives, many of us wonder whether technologies have taken over the world and assimilated humans, or if they are enabling us to transform ourselves into something new, that would liberate us from the tyranny of matter and the great oedipal drama that seems to give limited meaning to lives at a high price (and with low cost-effectiveness). In theory, technologies free us from attachments to bodies, to place, and to the so-called tedium of work. Yet until now, utopian thoughts about the fiction of technology have been outweighed by technological ties to consumerism, or to what some critics have called integrated world capitalism.[3]

Neither bathing in *fin-de-siècle* pathos nor engaging blindly and somewhat naively in celebration of waste, the chapters to follow critique present conditions but also speculate on the importance of new articulations between the humanities and the technosciences. They emphasize the necessity to think critically in an age based on easy consensus and mass-mediazation. As intellectuals, we must ask: What interpretations about the condition of everyday life can we offer that will not simply revert to other technologies of application that try to cast a grid over the objects they wish to control? Problems in natural and social ecology have shown that simple dialectics of progress no longer hold. Order may come out temporarily of disorder, and if human societies are in constant flux, so too is nature. How do we exit from a simple dialectic and enter into a changing world, yet in such a way that *becoming* remains a term reserved to humans *and/in* the world?

Technology is a mode of thinking, a special kind of technics that literally applies its own rules to itself and then becomes a self-fulfilling prophecy. The essays in this volume are aimed at problematizing these same effects of "self-application" that appear to equate technology with truth. They all ask, in other words, an archaic but always viable question: What is the position of technology in the humanities? An era has come undoubtedly where, again, the humanities cannot adequately deal with the world without assuming knowledge that can enable them to be in dialogue with new disciplines.

It can be said that the inverse also holds, and that applied scientists will have to find a "new alliance" with the world of metaphors in which the humanities find their innovation and renewal. This is what scientists such as Ilya Prigogine and Isabelle Stengers advocate.[4] Like some of the contributors to this volume, they introduce the concept of history into nature. They show that nature can no longer, for ideological purposes, be said to be inert and passive, or to be conquered and acted upon. In nature, concepts such as pattern and randomness, bifurcation in high-fluctuation conditions, and irreversible time may encourage scientists, faced with uncertainties, toward renewing the dialogue with humanities and introducing ethical dimensions in their disciplines. Ecological dilemmas prove that the world cannot be reduced to a scientific object, that it escapes total mastery, and that other ways—or *techné*—of approaching it must be essayed if humanity is to sustain its life.

In a shrinking world, where everything is at one's fingertips, populations have increased according to Malthus's predictions and are projected to double once again by the year 2050. Far from producing overabundance and ease, technologies that have fostered growth have extended social inequities. To the North-South axis of riches versus poverty, islands of riches are added everywhere, as also are gaping pockets of poverty. Violence escalates in forms of war and civil unrest. Ecological catastrophes multiply, from Three Mile Island to Chernobyl and Bangladesh, from massive deforestation to disappearing ozone layers, from epidemics to a slackening in food production. Evidence shows that technologies have not led humans toward any promised land.

In view of the grim prospect of the twenty-first century, we are compelled to ask how critics of culture, philosophers, and artists will deal with technologies. How do they contend with expansionist ideology, and the accelerated elimination of diversity and of singularities? How do they resist or act? Paul Virilio has argued that the changes societies are currently undergoing with electronic revolutions are as important as those introduced by monocular perspective in the Quattrocento. The introduction of the artificial perspective went hand in hand with a scientific revolution that brought about modern-day technology. It coincided with discovery and the invention of movable type, or automatic writing. Now, in a world where the notion of space has been completely changed through electronic simultaneity, where the computer appears to go faster than the human brain, or where "virtual reality" replaces "reality," how do philosophy, critical theory, or artistic practices deal with those shifts?

The essays included in this volume address these questions from different points of view.[5] Most reach back to Heidegger's article "The Question Concerning Technology," first published in 1953. There, the philosopher denounces modern, instrumental technology leading to control over nature as an arrestive enframing (*Gestell*) that tries to dominate and explicate. To this "enframed" use of technology, he opposes an ostensibly older *techné* that the Greeks called *poiesis*, a bringing forth, a setting-on-the-path toward revelation, truth, being, or essence. As he put it in the celebrated sentence, "The essence of technology is nothing technological."

Written forty years ago, just after a war that served as a testing ground for technologies whose developments and virtuality were well known in industrial and military sectors, Heidegger's article still appears at the center of the debates that call technologies into question. Reference to the article in this collection shows how it is necessary to go through, but also *beyond*, Heidegger, who thought in terms of domination of nature and of loss of humanness by way of technology, but neither of transformation of subjectivities nor of limits imposed by natural or social ecology. Taken up are reflections about how to distinguish between an instrumental technology and *techné*; how to navigate between the two or how to alter the distribution of terms; how to critique an ever-encroaching use in a market economy of technologies that lead to economic exploitation and a new kind of imperialist takeover; how genocide is a tributary term or subcategory of geocide. At stake is the singularity of the subject threatened with annihilation as much through technology as through the effects of mass media. The individual has a false feeling of autonomy and agency, while he or she is being drugged or manipulated into immobility. And perhaps autonomy was never a good way—a good *techné*—of asking the question. Does the rapid transformation of our space immobilize us (Virilio) or allow for other unknown ways of becoming beyond the Heideggerian dilemma (Guattari)? Emphasis on a becoming technological of the world may render obsolete or impossible ways of thinking in terms of sovereignty, be it on the level of individuals or of nations (Nancy). The same becoming technological that speeds things up only to slow them down describes a narcotic modernity bent on destruction and that falsely tries to scapegoat individuals who are designated addicts (Ronell).

Technologies have not just liberated us and given us progress. They have also devastated the environment and our living spaces. Any progressive cause can no longer think subjectivity outside a link through a rehistoricized nature (Conley). Our reading of the world through physics is linked to our psychic apparatus and our fantasized domination of the world (Brennan). What other relations to the world are there? Is the Heideggerian notion of *techné* as *poiesis* (Scheibler) adequate as a means of essaying other relations to the world? For the German Romantics, the artist occupied a position outside of society and was believed to be castigating bourgeois society, intent upon advocating new forms for the sake of economic gain. A formal avant-garde, based on experimentation in art, took a critical view of the surrounding world. It produced consciousness and meant to introduce social changes by advocating the creative autonomy of the subject. Such a project, it can be argued, failed through technologization and mass-mediazation as well as an increasing awareness of humans' interdependence among themselves and with nature that now includes the biosphere. What openings are available to us, what critical and artistic projects that would neither fall into the trap of the media (such as architecture) nor perpetuate the illusion of autonomy? Is formal experimentation as a political gesture still possible now that advanced technologies can reproduce anything at any time? Are we condemned to cynicism and to parodic repetition of forms of art, to kitschiness and simple technical performativity (Gaillard), or to a simulation that would reveal our inherent penchant for the death drive (Durham) but also a return to the subject? Yet simulation is outstripped in a way by recent experiments in cyberspace that transform our notion of mimesis. Cyberspace breaks down the barriers between virtual and real, and puts forward a new ensemble of questions. Is this adventure or nightmare (Hayles), or a way of going back to an analogical thinking (Moreiras) that is an end and a lost object of technology, as in poetry? Perhaps cyberspace can be enveloped in art and performance, and can be used to enable us paradoxically to break away from technological takeover all the while the written word is complicated by the image (Clancy).

With their variety of approaches, all of the contributors to this volume show how the necessity to rethink technologies is at the heart of our present and future existence. Technology continuously transforms our beliefs and our ways of living our bodies at unprecedented speeds. Exhilarating and frightening, absorbed by exploitative capitalist forces and leaving havoc on our habitat, technologies also promise other possibilities and beg to be rechanneled into more productive modes of singular and collective becoming.

Notes

1. Martin Heidegger, "The Question Concerning Technology," in *Basic Writings*, ed. David Farrell Krell (New York: Harper and Row, 1977), 284-317.

2. The expression "project occidental" is developed by Edouard Glissant in *Le discours antillais* (Paris: Seuil, 1980).

3. See Félix Guattari, *Les trois écologies* (Paris: Galilée, 1990).

4. Ilya Prigogine and Isabelle Stengers, *La nouvelle alliance* (Paris: Gallimard, 1979); translated as *Order out of Chaos* (foreword by Alvin Toffler) (Boulder, Colo.: New Science Library, 1984).

5. These different points of view are all connected to the transforming notion of community that was the subject of the first L. P. Irvin Colloquium, "Community at Loose Ends," held at Miami University in October 1988—and published by the University of Minnesota Press in 1991. The second Irvin Colloquium addressed these issues under the title "Questioning Technologies." Technology was the theme for the entire year 1990-91 upon which the papers of those who could not attend the October colloquium were also based. As is the custom, participants were sent in advance a number of questions concerning technology that they were asked to address in their papers.

Acknowledgments

The Miami Theory Collective would like to thank the following for their participation and their contribution to this project: Jesse Dickson, Alain Gabon, Giuliana Menozzi, Carl Pletsch, Sylvie Blum-Reid, Peter Rose, Edward Tomarken, and Michèle Vialet. We would also like to thank Biodun Iginla, for his continued encouragement and support, and Steven Ungar, for his patient readings and valuable suggestions. Our thanks go also to Karlene Off and Collene McDevitt for their assistance. We are especially grateful to Donna Cheshire, without whose initiative, hard work, and wry humor this volume could not have been completed.

Above all, we would like to thank Dr. William Stitt and Mrs. Evelyn Stitt, whose generous contribution made the Technology Colloquium possible. To them this volume is dedicated.

ACKNOWLEDGMENTS

PART I
Questioning Technologies

Chapter 1

The Third Interval: A Critical Transition

Paul Virilio

We know about critical *mass*, critical *instant*, and critical *climate*: we hear less often about critical space. There is no easy reason for this, unless perhaps it is because we have not yet assimilated *relativity*, the very notion of space-time. And yet space, or critical extension, has become ubiquitous, because of the acceleration of "means of communication" that *collapse the Atlantic* (the Concorde), *reduce France to a square of an hour and a half on each side* (the airbus), or, yet again, tell us that the high-speed train (TGV) *wins time over time*. These different slogans from the world of publicity indicate exactly how much we inherit old ideas of geophysical space; these advertisements also tell us, to be sure, that we are their innocent victims. Today we are beginning to realize that systems of telecommunication do not merely confine *extension*, but that, in the transmission of messages and images, they also eradicate *duration* or delay.

In the shift from the revolution of modes of transportation in the nineteenth century to the revolution of electronic communication in the twentieth century, there emerge a mutation and a commutation that affect public and domestic space so strongly that we are hard put to determine what its reality may be. When technologies of *telemarketing* replace those of the classical era of *television*, we begin to witness how the premises of an urbanization of *real time* follow on the heels of the premises of an urbanization of a real *space*. Because of interactive teletechnologies (the teleport), this abrupt transfer of technology moves from the arrangement of the infrastructures of real space (maritime ports, railway stations, airports) to the control of the environment in real time. Critical dimensions are also being renewed.

The question of the *real moment* of instantaneous telemarketing is effectively refashioning philosophical and political issues that traditionally had been based on notions of *Atopia* and *Utopia*. The shift is being made for the advancement of what has already been called *Teletopia*, which carries manifold paradoxes that take, for example, the following form: ''Reach out and touch someone,'' or even ''to be telepresent,'' meaning to be here and elsewhere *at the same time*. This so-called real time is essentially nothing other than a real space-time, since different events surely take ''place'' even if, finally, this place constitutes that of the no-place of teletopical technologies (such as the interface of human and machine, a regime or nodal point of teletransmissions).

Immediate telesales, instant telepresence: thanks to new procedures of telediffusion or of teletransmission, *action*, or the fabled ''televised action at a distance'' that the telecommander effectuates, is now facilitated by the perfected use of electromagnetics and by the radio-electric views of what has lately been called electro-optics. One by one, the perceptive faculties of an individual's body are transferred to machines, or instruments that record images and sound; more recently, the transfer is made to receivers, to sensors, and to other detectors that can replace absence of tactility over distance. A general use of telecommands is on the verge of achieving permanent telesurveillance. What is becoming critical here is no longer the concept of three spatial dimensions, but a fourth, temporal dimension—in other words, that of the present itself. As we shall see below, ''real time'' is not opposed—as many experts in electronics claim—to ''deferred time,'' but only to *present* time.

The painter Paul Klee expressed the point exceptionally well when he noted, ''Defining the present in isolation is tantamount to murdering it.''[1] This is what technologies of real time are achieving. They kill ''present'' time by isolating it from its presence *here and now* for the sake of another commutative space that is no longer composed of our ''concrete presence'' in the world, but of a ''discrete telepresence'' whose enigma remains forever intact. How can we fail to understand to what degree these radio-technologies (based on the digital signal, the video signal, and the radio signal) will soon overturn not only the nature of human environment and its *territorial body*, but also the individual environment and its *animal body*, since the development of territorial space by means of heavy *material* machinery (roads, railways, and so on) is now giving way to an almost *immaterial* control of the environment (satellites, fiber-optic cables) that is connected to the terminal body of the men and women, interactive beings who are at once emitters and receivers?

Clearly the urbanization of real time entails first of all the urbanization of ''one's own body,'' which is plugged into various interfaces (computer keyboards, cathode screens, and soon gloves or cyberclothing), prostheses that turn the overequipped, healthy (or ''valid'') individual into the virtual equivalent of the well-equipped invalid. If the revolution of modes of transportation of the last century had witnessed the emergence and progressive popularization of the dynamic automotive vehicle (train, motorcycle, car, airplane), the current electronic revolution is now, in its turn, blueprinting the plan for the innovation of the ultimate vehicle, the *static audiovisual vehicle*, in other words, the coming of a behavioral inertia of the receiver-sender, or the passage from this fabled "retinal suspension," on which the optical illusion of cinematic projection was based, to the "bodily suspension" of the "plugged-in human being." This becomes the condition of possibility of a sudden mobilization of the illusion of the world, of an *entire* world, that is telepresent at every moment. The very body of the connected witness happens to be the ultimate urban territory—a folding back over the animal body of social organization and of a conditioning previously limited to the core of the old city. In bodily terms, it resembles the core of the old familial "hearth."

Thus we are better able to perceive the decline of the unity of a demography. After an expanse of time the extended family turned into the nuclear family, which has now become the single-parent family. Individuality or individualism was thus not so much the fact of a liberation of social practice as the product of the evolution of techniques of the development of public or private space. If cities are growing and sprawling at unforeseen rates, so then the familial unit is shrinking and becoming a tributary force. Given that we are witnessing supersaturated conditions in the concentrations of megalopolitan populations (Mexico City, Tokyo, Los Angeles) that are the result of an increased economic speed, it now seems appropriate to reconsider the notions of acceleration and deceleration (what physicists call positive and negative speeds) and, no less, what is less evident, in *real* speed and *virtual* speed (the rapidity of what happens unexpectedly, such as an urban crisis, or an accident) to grasp better the importance of the "critical transition" of which we are now the powerless witnesses.

We would do well to recall that speed is not a phenomenon but a relation among phenomena, in other words, *relativity itself*, whence the importance of the constancy of the speed of light not only in physics or in astrophysics, but also in our everyday lives. It is experienced as soon as we move, beyond the paradigm of public transport, into that of the organization and *electromagnetic conditioning* of territorial space. Such is what is implied by revolutions in "transmission" or automation, of environmental control in real time that has since replaced traditional ways of living in territorial space. As a result, speed is not used solely to make travel more effective. It is used above all to see, to hear, to perceive, and, thus, to conceive more intensely the present world. In the future, speed will be used more and more to *act over distance*, beyond the sphere of influence of the human body and its behavioral biotechnology.

The Interval of Light

How can we account for this situation? It is necessary to introduce the specter of a new kind of interval, the *interval of light* (or zero-sign). In fact, in relativity the revolution of this third ''interval'' is in itself a sort of imperceptible cultural revolution. If the interval of *Time* (a positive sign) and the interval of *Space* (a negative sign) have given impetus to the geography and the history of the world through geometrical measurement of agrarian space (allotment into parcels of land) and urban areas (cadastral surveys), the organization of the calendar and measurement of time (clocks and watches) have also presided over a vast political and chronological regulation of human societies. The sudden emergence of an interval of the third type thus signals that we are undergoing an abrupt qualitative shift, a profound mutation of the relations that as humans we are keeping with our living environment. *Time* (duration) and *Space* (extension) are now inconceivable without *Light* (absolute speed), the cosmological constant of the speed of light, an absolute philosophical contingency, according to Einstein, that follows the absolute character that until then Newton and his predecessors had ascribed to space and time.

Since the beginning of this century, the absolute limit of the speed of light has, as it were, *enlightened* space and time together. We are therefore no longer dealing so much with light that illuminates things (the object, the subject, and travel) as with the constant character of its absolute speed, which conditions the phenomenal apperception of the world's duration and extension.[2] We do well to heed the physicist who speaks of the logic of particles: ''A representation is defined by a sum of observables that are flickering back and forth.''[3] The macroscopic logic of the techniques of *real time* could not better describe the macroscopic logic of this sudden ''teletopical commutation'' that perfects what until now had been the fundamentally ''topical'' quality of the old human city.

Thus both the urban geographer and the political scientist find themselves torn between the permanent necessities of the organization and construction of real space, with all of its basic problems, including geometrical and geographical constrictions about what is central versus what is peripheral, and new constraints of the management of this real time of immediacy and ubiquity, with its ''protocol of access,'' its ''transmission of bundles,'' its ''viruses,'' and the chronogeographical constraints of nodal and interconnected networks. An extended time works in the direction of the topical and architectonic interval (the high-rise building), and a short, ultrabrief, even inexistent time in the direction of the teletopical interface (the network). How can this dilemma be resolved? How can these fundamentally spatiotemporal and relativistic problems be formulated?

When we now witness the aftershocks of international financial disasters in view of the damages of instantaneous automation of stock futures and junk bonds, or this notorious trading program that is responsible for the acceleration of economic disorder—such as the electronic crash of October 1987 and the crash that was barely missed in October 1989—we put our finger on the difficulties of our current situation.

Critical transition is thus not a gratuitous expression: behind this vocable there lurks a real crisis of the temporal dimension of *immediate action*. After the crisis of "integral" spatial dimensions, which give increased importance to "fractional" dimensions, we might be witnessing, in short, the crisis of the temporal dimension of the present moment. If time-light (or, better, the time of the speed of light) now serves as an absolute standard for both immediate marketing and instantaneous telemarketing, then intensive duration of the "the real moment" now replaces duration. Thus the extensive time of history is relatively subject to control, and can include this long-term duration, what used to comprise at once the past, the present, and the future. In effect, what we might call a temporal commutation, an "alternation" or "flickering" that is also related to a sort of *commotion* of present duration, an accident of a so-called real instant, is suddenly disconnected from its site of origin or inscription, from its here and now, for the sake of an electronic dazzle (that is at once electro-optical, electro-acoustical, and electro-tactile) where telecommanding, the so-called tact at a distance, would bring to completion the former technique of telesurveillance of what is kept afar, or beyond our grasp.

If, as Epicurus says, *time is the accident of accidents*, with these teletechnologies of generalized interactivity we begin to move toward the era of the accident of the present, the fabled telepresence over distance that amounts to nothing more than the sudden catastrophe of the reality of this *present instant* that constitutes our only mode of entry into duration, but also—and everyone has been aware of the fact since Einstein—our only entry into the extension of the real world. Henceforth the "real" time of telecommunications will probably refer no longer solely to "deferred" time, to feedback, or to time lags, but also to an *outer chronology*. Whence my constantly reiterated point about replacing what is *chronological* (before, during, after) with what is *dromological* or, if another formula fits better, the *chronoscopical* (underexposed, exposed, and overexposed). In effect, the interval of light (the interface) supplanting henceforth those of Space and of Time, the notion of *exposure* replaces, in its turn (whether we like it or not), that of *succession* in terms of present duration and that of *extension* in immediate space.

Thus the speed of exposure of time-light should allow us to reinterpret the "present" or this "real instant" that is (lest we forget) the space-time of a very real action facilitated by electronic machines. Soon it will be facilitated by photonic apparatus, that is, by the absolute capacities of electromagnetic waves and of quanta of light, a limit and a milestone for access to the reality of the perceptible world (here I am thinking of what astrophysicists call the cone of light).

The dilemma that these teletopical techniques are currently posing thus becomes a major problem for city planners, because the urban mapping of real time fostered by the recent revolution of communications leads to a radical *inversion* in the order of the movement of travel and physical displacement. Indeed, if *control at a distance* encourages the increasing suppression of material infrastructures that had equipped territory for the sake of the basically immaterial character of the airwaves of telesurveillance and instantaneous telecommanding, it is because *travel* and its adjuncts undergo an obvious mutation and commutation. Where, in the past, physical displacement from one point to another presupposed a departure, a voyage, and an arrival, more than a century ago revolutions in modes of transport had already set in place a liquidation of delay and oriented the very nature of travel (on foot, on horseback, and in a car!) toward the arrival at a final point that remained, however, a *restricted arrival* by virtue of the very duration of the voyage.

Currently, with the revolution of instantaneous transmissions, we are witnessing the beginnings of a type of *general arrival* in which everything arrives so quickly that departure becomes unnecessary. The liquidation of "travel" (that is, the interval of space and time) in the nineteenth century is now replaced, at the end of the twentieth century, by the elimination of "departure." The journey, with its succession of events that had previously defined its character, is lost in favor of the sole immediacy of "arrival." A type of *general arrival* can explain today the incredible innovation of the *static vehicle* that is not only audiovisual but *tactile* and *interactive* in nature (that is, radio-active, opto-active, and interactive). Such is Scott Fisher's new *costume of givens*. NASA is sponsoring work to produce an apparatus for the human body that will transmit actions and sensations by receivers and sensors, its *presence in distance* (and this means any expanse of distance, because the NASA project should permit the total telemanipulation of a double robotics on the ground of the planet Mars) thus achieving an effective "telepresence" of the individual *in both places at the same time*. The doubling of the personality of the manipulator would be achieved by the "vehicle" that makes up his or her instantaneous interactive vector.

One of Paul Klee's ultimate visionary remarks can be recalled: "For the spectator most activity is temporal."[4] But what can be said of the *teleactors'* interactivities? Are their experiences, like those of the classical *telespectator*, no less spatial than temporal? Consigned to inertia, interactive beings transfer their natural capacity of movement and travel into probes, into detectors that inform their users immediately about distant realities, but to the detriment of their own sensory faculties of reality. Examples include the para- or tetraplegic who is able to teleguide his or her environment or inhabited space as a model of these *domotics* and the "smart high rises" that will soon respond to every one of our velleities. Thus the *mobile* human who had become *automobile* will now become *motile*, willfully limiting his or her bodily sphere of influence to a few simple gestures, to the emission—or zapping—of several signs.

The critical situation of life that numerous motorized handicapped citizens have endured is developing, by the force of things (*the critical force of technical things*), into the paradigm of the human being of the future, that is, the inhabitant of the future teletopical city. It will be a *metacity* rife with social deregulation, whose transpolitical aspect has already begun to emerge now and again in a number of minor incidents or major accidents that still remain unexplained.

The Critical Transition

How can we best grasp this transitional situation, in other words, what physicists call this ''transitional phase''? One of Nicholas of Cusa's early philosophical analyses comes to mind:

> Although the accident disappears when a substance is removed, the accident is not, however, a total absence. If it perishes, it is because it is part of its very accidental nature to be attached to another reality. The accident brings so much to substance that although it receives its being from substance alone, *substance cannot exist in the absence of all accident*.[5]

Today, as we have seen above, the question of the ''accident'' has shifted from the space of matter to the time of light. An accident is first of all an *accident of transmission* of the absolute speed of electromagnetic waves, a speed that hereafter allows us not only to hear and to see, as we had in the era of the telephone, the radio, or television, but now to act over greater distances: whence the necessity of a third type of interval (of a zero-degree sign). With it we can grasp *the place of a nonplace* of a teleaction no longer associated with the here and now of immediate or local action. An ''accident of transmission'' in the world of interactivity thus opens not only onto a transmission of technology between communication in deferred time and commutation in real time, but most of all onto a political transference that calls into question the dilemmas that are rightly at the center of our time, especially those concerning what is *public* and what is *service*.

Effectively, what remains of the notion of ''service'' when we are likewise ''served'' and thus subdued [*asservi*]? What remains of the notion of things ''public'' when public *images* (in real time) are more important than public *space*? Already the notion of ''public transport'' is slowly giving way to that of a *connected travel*, in which continuity wins over discontinuity. What does our being *plugged in at home* have to say about ''domotic'' domesticity in high-rise buildings like those in an intelligent and interactive city such as Kawasaki? By eliminating former geopolitics, the crisis of the notion of physical dimension thus is colliding head-on with the politics and administration of public service. Thus, if the classical interval gives way to interfacing, politics moves, in turn, into *present time* alone. The question no longer entails relations of what is global in respect to what is local, or what is transnational and what is national, but above all concerns this sudden ''temporal commutation'' in whose flickerings disappear not only the difference of inside and outside and the expanse of political territories, but also the ''before'' and the ''after'' of duration and history, for the sake of a *real instant* over which, finally, no one has control. To be convinced of this shift we need only observe today's inextricable problems of geostrategy in view of the impossibility of clearly distinguishing *offense* from *defense*. Instantaneous and multipolar strategy has been deployed in what military experts call ''preemptive'' strikes!

Thus the archaic ''tyranny of distances'' between people who have been geographically scattered increasingly gives way to this ''tyranny of real time'' that is not merely a matter—as optimists might claim—for travel agencies, but especially for employment agencies, because the more the speed of commerce grows, the more unemployment becomes globally massive. Since the nineteenth century, the muscular force of the human being is literally ''laid off'' when automation of the ''machine tool'' is employed. Then, with the recent growth of computers, ''transmission machines,'' comes the laying off or ultimate shutdown of human memory and conscience. Automation of postindustrial production is coupled with the automation of perception and then with this *attended* conception favored by the marketplace of systems analysis while future developments are sought in cybernetics. Thus, the gain of *real* time over *deferred* time is equivalent to being placed in an efficient procedure that physically eliminates the ''object'' and ''subject'' for the exclusive advantage of a journey, but the journey [*trajet*], because it lacks a trajectory, is fundamentally out of control. Thus the *interface* in real time definitely replaces the *interval* that had formerly constructed and organized the history and geography of our societies, leading to an obvious culture of paradox, in which everything *arrives* without there being any need either to *travel* or to *leave* in the slightest physical sense.

Behind this critical transition, how can we fail to wonder about the future conditioning of the human environment? If the revolution of transportation in the nineteenth century had already prompted a change in the surface urban territory on the whole of the European continent, the current revolution of interactive transmissions is, in its turn, promoting an alteration of urban environment. ''Images'' win over the ''things'' they are said to represent: the city of the past slowly becomes a paradoxical agglomeration in which relations of immediate proximity give way to interrelations over distances. In fact, the paradoxes of acceleration are frequent and disturbing. One—the first—of them runs thus: when things ''far'' are brought into immediate proximity, those that are proportionately ''near''—such as our friends, kin, neighbors—turn what is proximate—family, work, or neighborhood—into a foreign, if not inimical, space. This inversion of social practices can already be seen in the urban planning of modes of communication (maritime port, railway station, airport) and is underscored and radicalized through new means of telecommunication (the teleport).

Once again we thus observe still another inversion of tendencies. Where motorized transportation and information had prompted a *general mobilization* of populations swept up in the exodus of labor (and then of leisure), modes of instantaneous transmission prompt the inverse, that of a *growing inertia*. Television and, especially, teleaction, no longer require human mobility, but merely a local motility. Telemarketing, tele-employment, fax work, bit-net, and e-mail transmissions at home, in apartments, or in cabled high rises—these might be called cocooning: an urbanization of real time thus follows the urbanization of real space. The shift is ultimately felt in the very body of every city dweller, as a *terminal citizen* who will soon be equipped with interactive prostheses whose pathological model is that of the "motorized handicapped," equipped so that he or she can control the domestic environment without undergoing any physical displacement. We have before us the catastrophic figure of an individual who has lost, along with his or her natural mobility, any immediate means of intervening in the environment. The fate of the individual is handed over, for better or for worse, to the capacities of receivers, sensors, and other long-range detectors that turn the person into a being subjected to the machines with which, they say, he or she is "in dialogue!"[6]

To be a subject or to be subjected? That is the question. Former public services will in all likelihood be replaced by a domestic enslavement for which "domotics" might be the perfect outcome. It would be equivalent to the achievement of a domiciliary inertia, where a generalization of techniques of "environmental control" would end up with behavioral isolation and reinforce cities with the very *insularities* that have always threatened them, such that the distinction between the "island retreat" and the "ghetto" might become increasingly precarious.

Furthermore, and for some unexplainable reason, the international colloquium on the handicapped that recently took place at Dunkirk offers numerous parallels with the critical situation that I have sketched in the paragraphs above. It appears as if the recent technical and economic imperatives insert *continuities* and *networks* in the place of *discontinuities*, where there existed an amalgam or mix of different types of urban mobilities. Whence the idea, described above, of a common public transit is replaced by that of a more pervasive *chain of displacement*. We can thus heed the generous conclusion François Mitterrand stated at the end of the Dunkirk symposium: "Cities will have to be adapted to their citizens, and not the other way around. We must open the city to handicapped citizens. I demand that a global politics for the handicapped become a strong axis of social Europe." If every one of us is obviously in agreement about the inalienable right that the handicapped person has to live as others do and therefore *with others*, it is no less revealing to note the similarities that now exist between the reduced mobility of the equipped invalid and the growing inertia of the over-equipped, "valid" human population. As if the revolutions in transmission of information led to an identical conclusion, whatever may be the condition of the patient's body, the *terminal citizen* of a teletopical city is on the way toward its accelerated formation.

The destruction of the Berlin Wall? That has been accomplished. The future of a united Germany? The answer is clear. The abolition of borders dividing nations in Western Europe is announced for 1993. What remains to be abolished—and urgently—can only be space and time. As we have just seen, the task is being accomplished. At the end of our century not much will remain of this planet that is not only polluted and impoverished, but also shrunken and reduced to nothing by the teletechnologies of generalized interactivity.

Translated by Tom Conley

Notes

1. Paul Klee, *Théorie de l'art moderne* (Paris: Gonthier, 1963).
2. [The triad described in the parentheses reads "*l'objet, le sujet, le trajet*," such that "travel" or "journey," the third term, bears strong graphic and vocal resemblance to the object and the subject.—Trans.]
3. G. Cohen Tannoudji and M. Spiro, *La matière-espace-temps* (Paris: Fayard, 1986).
4. Klee, *Théorie de l'art moderne*.
5. Cited by Guiseppe Bufo, in *Nicolas de Cues* (Paris: Seghers, 1964).
6. See Paul Virilio, *L'inertie polaire* (Paris: Christian Bourgois, 1990).

Chapter 2

Machinic Heterogenesis

Félix Guattari

Machinism

Although machines are usually treated as a subheading of "technics," I have long thought that it was the problematic of technics that remained dependent on the questions posed by machines. "Machinism" is an object of fascination, sometimes of delirium. There exists a whole historical "bestiary" of things relating to machines. The relation between human and machine has been a source of reflection since the beginning of philosophy. Aristotle considers that the goal of *techné* is to create what nature finds it impossible to achieve so that *techné* sets itself up between nature and humanity as a creative mediation. But the status of this "intercession" is a source of ambiguity. While mechanistic conceptions of the machine rob it of anything that can differentiate it from a simple construction *partes extra partes*, vitalist conceptions assimilate it to living beings, unless the living beings are assimilated to the machine. This was the path taken by Norbert Wiener as he opened up the cybernetic perspective in *Cybernetics*.[1] On the other hand, more recent systemist conceptions reserve the category of autopoiesis (or self-production) for living machines (in Francisco Varela's *Autonomie et connaissance*),[2] whereas an older Heideggerian mode of philosophy entrusts *techné*, in its opposition to modern technicity, with the mission of "unveiling the truth," thus setting it solidly on an ontological pedestal—on a *Grund*—that compromises its definition as a process of opening. It is by navigating between these two obstacles that we will attempt to discern the thresholds of ontological intensity that will allow us to grasp "machinism" [*le machinisme*] all of a piece in its various forms, be they technical, social, semiotic, or axiological. With respect to each type of machine, the question will be raised not of its vital autonomy according to an animal model, but of its specific enunciative consistency.

The first type of machine that comes to mind is that of material assemblages [*dispositifs*], put together artificially by the human hand and by the intermediary of other machines, according to diagrammatic schemas whose end is the production of effects, of products, or of particular services. From the outset, through this artificial montage and its teleology [*finalisation*] it becomes necessary to go beyond the delimitation of machines in the strict sense to include the functional ensemble that associates them with humankind through multiple components:

material and energy components;
semiotic components that are diagrammatic and algorithmic;
social components relative to the search, the formation, the organization of work, the ergonomics, the circulation, and the distribution of goods and services produced;
the organ, nerve impulse, and humoral components of the human body;
individual and collective information and mental representation;
investments by "desiring machines" producing a subjectivity in adjacency with its components;
abstract machines setting themselves up transversally to the machinic, cognitive, affective, and social levels considered above.

In the context of such a functional ensemble, which henceforth will be qualified as *machinic ordering* [*agencement machinique*], the utensils, the instruments, the simplest tools, and, as we shall see, the slightest structured parts of a machinery will acquire the status of a protomachine. Let us deconstruct, for example, a hammer by removing its handle. It remains a hammer but in a "mutilated" state. The "head" of the hammer, another zoomorphic metaphor, can be reduced by fusion. It will then cross the threshold of formal consistency, causing it to lose its form, its machinic gestalt, which works on a technological as well as on an imaginary level (as, for example, when we evoke the obsolete memory of the hammer and sickle). From then on, we are confronted with nothing more than a metallic mass that has been returned to its smooth state—to deterritorialization—preceding its entrance into that mechanical form. But we will not settle for this experiment, similar to Descartes's experiment with a piece of wax. In effect, we can move in the opposite direction of this deconstruction and its limit threshold, toward the association of the hammer and the arm, the nail, the anvil, which maintain among each other relationships that we can call syntagmatic. Their collective dance even expands to include the defunct corporation of blacksmiths, the sinister epoch of the old iron mines, ancestral use of iron-rimmed wheels. As Leroi-Gourhan pointed out, the technological object is nothing outside of the technological ensemble to which it belongs. But is it any different with sophisticated machines such as robots, which we suspect—probably with good reason—will soon be engendered exclusively by other robots in a gestation involving virtually no human action until some glitch requires our residual, direct intervention? But doesn't all that sound like a kind of dated science fiction? In order to acquire more and more life, machines require more and more abstract human vitality as they make their way along their evolutive phyla. Thus, conception by computer—expert systems and artificial intelligence—gives us back at least as much as it takes away from thought, because in the final analysis it only subtracts inertial schemas. Computer-assisted forms of thought are thus mutant and arise from other kinds of music, from other universes of reference.

It is thus impossible to refuse human thought its part in the essence of machinism. But how long can we continue to characterize the thought put to work here as human? Doesn't technicoscientific thought emerge from a certain type of mental and semiotic machinism? Here it becomes necessary to establish a distinction between, on the one hand, semiologies producing significations that are the common currency of social groups and, on the other, asignifying semiotics that, despite the significations they can foster, manipulate figures of expression that work as diagrammatic machines in direct contact with technical-experimental configurations. Semiologies of signification play on distinctive oppositions of a phonemic or scriptural order that transcribe enunciations [*énoncés*] into expressive materials that signify. The structuralists liked to make the Signifier a unifying category for all expressive economies of whatever order, be it language, icon, gesture, urbanism, or cinema. They postulated a general translatability able to signify all forms of discursivity. But in doing that, did they not miss the mark of a machinistic autopoiesis that does not derive from repetition or from mimesis of significations and their figures of expression, but that is linked instead to the emergence of meaning and of effects that are no less singular for being indefinitely reproducible?

Ontological Reconversions

This autopoeitic nexus of the machine is what wrests it from structure. Structural retroactions, their input and output, are called upon to function according to a principle of eternal return; they are inhabited by a desire for eternity. The machine, on the contrary, is haunted by a desire for abolition. Its emergence is accompanied by breakdown, by catastrophe, by the threat of death. Later on we will have to examine the different relations of alterity thus developed, relations that constitute differences from structure and its homeomorphic principle. The principle of difference proper to machinistic autopoiesis is based on disequilibrium, on prospecting for virtual universes far from equilibrium. And it is not just a question of a formal rupture of equilibrium, but a radical ontological reconversion. And that is what definitively denies any far-reaching importance to the concept of Signifier. The various mutations of ontological referent that shunt us from the universe of molecular chemistry to the universe of biological chemistry, or from the world of acoustics to the world of polyphonic and harmonic music, are not brought about by the same signifying entities. Of course, lines of signifying decipherability, composed as they are of discrete figures subject to being converted into binary oppositions, syntagmatic and paradigmatic chains, can be linked from one universe to another so as to give the illusion that all phenomenological regions are woven together in the same fabric. But things change completely when we turn to the texture of these universes of reference, which are, each time, singularized by a specific constellation of expressive intensities, given through a pathic relationship, and delivering irreducibly heterogeneous ontological consistencies. We thus discover as many types of deterritorialization as we do characteristics of expressive matter. The signifying articulation that looms above them—in its superb indifference and neutrality—is unable to impose itself upon machine intensities as a relation of immanence. In other words, it cannot preside over what constitutes the nondiscursive and self-enunciating nexus of the machine. The diverse modalities of machine autopoiesis essentially escape from signifying mediation and refuse to submit to any general syntax describing the procedures of deterritorialization. No binary couple such as being/entity [*être/étant*], being/nothingness, being/other can claim to be the "binary digit" of ontology. Machinic propositions escape the ordinary game of energetic/spatial/temporal discursivity. Even so, there nevertheless exists an ontological "transversality." What happens at a particle/cosmic level is not without relationship to what happens at the level of the socius or the human soul, but not according to universal harmonics of a platonic nature (as in "The Sophist"). The composition of deterritorializing intensities is incarnated in machines that are abstract and singularized, machines that have the effect of rendering things irreversible, heterogeneous, and necessary. On this score, the Lacanian signifier is doubly inadequate. It is too abstract in that it renders too easily translatable the materials of heterogeneous expressions; it falls short of ontological heterogenesis; it gratuitously renders uniform and syntactic the diverse regions of being. At the same time, it is not abstract enough because it is incapable of accounting for the specificity of these autopoietic nexes, to which we must now return.

Autopoietic Nexus

Francisco Varela characterizes a machine as "the ensemble of the interrelations of its components, independent of the components themselves."[3] The organization of a machine thus has nothing to do with its materiality. From there Varela goes on to distinguish two types of machines: allopoietic machines, which produce something besides themselves, and autopoietic machines, which continually engender and specify their own organization and their own limits. They carry out an incessant process of replacing their components because they are subject to external perturbations for which they are constantly forced to compensate. In fact, Varela reserves the qualification "autopoietic" for the biological domain. Social systems, technical machines, crystalline systems, and so forth are excluded from the category. That is the sense of his distinction between allopoiesis and autopoiesis. But autopoiesis, which thus encompasses only autonomous, individuated, and unitary entities that escape relations of input and output, lacks characteristics essential to living organisms, such as being born, dying, and surviving through genetic phyla. It seems to me, however, that autopoiesis deserves to be rethought in relation to entities that are evolutive and collective, and that sustain diverse kinds of relations of alterity, rather than being implacably closed in upon themselves. Thus institutions, like technical machines, which, in appearance, depend on allopoiesis, become ipso facto autopoietic when they are seen in the framework of machinic orderings that they constitute along with human beings. We can thus envision autopoiesis under the heading of an ontogenesis and phylogenesis specific to a mecanosphere that superimposes itself on the biosphere.

The phylogenetic evolution of machinism can be construed, at a first level, in the fact that machines arise by "generations"; they supersede each other as they become obsolete. The filiation of past generations is continued into the future by lines of virtuality and by their implied genealogical descendancy [*arbres d'implication*]. But we are not talking about a univocal historical causality. Evolutive lineages present themselves as rhizomes; datings are not synchronic but heterochronic. For example, the industrial ascendancy of steam engines took place centuries after the Chinese empire had used them as children's toys. In fact, these evolutive rhizomes traverse technical civilizations by blocks. A technological mutation can know periods of long stagnation or regression, but it is rare for it not to resurface at a later time. That is particularly clear with technological innovations of a military nature, which frequently punctuate large-scale historical sequences that they stamp with a seal of irreversibility, wiping out empires in favor of new geopolitical configurations. But, I repeat, the same was already true of the humblest instruments, utensils, and tools that are part of the same phylogenesis. One could, for example, mount an exposition on the subject of the evolution of the hammer since the stone age, and produce conjectures about what it might become in the context of new materials and new technologies. The hammer we buy today at the hardware store is, in some ways, "appropriated" from a phylogenetic lineage with virtual possibilities for the future that are undefined.

The movement of history is singularized at the crossroads of heterogeneous machinic universes, of differing dimension, of foreign ontological texture, with radical innovations, with benchmarks of ancestral machinisms previously forgotten and then reactivated. The neolithic machine associates, among other components, the machine of spoken language, the machines of cut stone, the agrarian machines founded on the selection of seeds and a protovillage economy. The scriptural machine, on the other hand, will see its emergence only with the birth of urban megamachines (compare Lewis Mumford) correlated to the implantation of archaic empires. In a parallel fashion, great nomadic machines will be constituted from the collusion between the metallurgical machine and new war machines. As for the great capitalistic machines, their basic machinisms were proliferative: first urban, then royal state machines, commercial and banking machines, navigational machines, monotheistic religious machines, deterritorialized musical and plastic machines, scientific and technical machines, and so forth.

The question of the reproducibility of machines on an ontogenetic level is more complex. The maintenance of a machine is never fail-safe for the presumed duration of its life. Its functional identity is never absolutely guaranteed. Wear and tear, precariousness, breakdowns, and entropy, as well as normal functioning, require a certain renovation of a machine's material, energetic, and informational components, the last of which is susceptible to disappearing in "noise." At the same time, maintenance of the consistency of machinic ordering requires that the quotient of human gesture and intelligence that figures in its composition must also be renewed. Man-machine alterity is thus inextricably linked to a machine-machine alterity that plays itself out in relations of complementarity or agonistics (between war machines) or else in the relations of parts or assemblages [*pièces ou dispositifs*]. In fact, wear and tear, accident, death, and resurrection of a machine in a new "example" or model are part of its destiny and can be foregrounded as the essence of certain aesthetic machines (Cesar's "compressions," "Metamechanics," happening machines, Jean Tinguely's machines of delirium). The reproducibility of machines is thus not a pure, programmed repetition. Its rhythms of rupture and fusion, which disconnect its model from all grounding, introduce a certain quotient of difference that is as ontogenetic as it is phylogenetic. On the occasion of these phases of transformation into diagrams, into abstract and disincarnated machines, the "soul supplement" of the machine nexus is granted its difference relative to simple material agglomerate. A pile of stones is not a machine, whereas a wall is already a static protomachine, manifesting virtual polarities, an inside and an outside, a high and a low, a right and a left. These diagrammatic virtualities lead us away from Varela's characterization of machinic autopoiesis as unitary individuation, without input or output, and prompts us to emphasize a more collective machinism, without delimited unity and whose autonomy meshes with diverse bases for alterity. The reproducibility of the technical machine, unlike that of living beings, does not rely upon perfectly circumscribed sequences of coding in a territorialized genome. Each technological machine has indeed its own plans of conception and assemblage, but, on the one hand, these are not conflated with the machine, and on the other hand, they get sent from one machine to another so as to constitute a diagrammatic rhizome that tends to cover the mecanosphere globally. The relations of technological machines among themselves, and adjustments of their respective parts, presuppose a formal serialization and a certain loss of their singularity—more so than in living machines—that is correlative to the distance assumed between the machine (manifested in the coordinates of energy/space/time) and the diagrammatic machine that develops in coordinates that are more numerous and more deterritorialized.

This deterritorializing distance and this loss of singularity must be attributed to a stronger smoothing out of the materials constitutive of the technical machine. Of course, the irregularities particular to these materials can never be completely smoothed out, but they should not interfere in the "freeplay" [*jeu*] of the machine unless required to by its diagrammatic function. Using a seemingly simple machinic ordering [*agencement machinique*], let us look closer at the couple formed by a lock and its key, at these two aspects of machinic separation and smoothing out. Two types of form, characterized by heterogeneous ontological textures, are at work here:

1. Materialized forms, which are contingent, concrete, and discrete, forms whose singularity is closed on itself, incarnated in profile F(L) of the lock and profile F(K) of the key. F(L) and F(K) never coincide completely. They evolve in the course of time as a result of wear and oxidation. But both are obliged to remain within the framework of a delimiting standard deviation beyond which the key would no longer be operational.
2. Diagrammatic, "formal" forms, subsumed by this standard deviation, which are presented as a continuum including the whole gamut of profiles F(K) and F(L) compatible with the effective unlatching of the lock.

We notice right away that the effect, the possible act of opening the lock, is located altogether in the second (diagrammatic) type of form. Although they are graduated according to the most restricted possible standard deviation, these diagrammatic forms appear in infinite number. In fact, we are dealing with an integral of forms F(K) and F(L).

This integral, "infinitary" form doubles and smooths out the contingent forms F(K) and F(L), which have machinic value only to the extent that they belong to it. A bridge is thus established "over" the authorized concrete forms. This is the operation that I am qualifying as deterritorialized smoothing out, an operation that has just as much bearing upon the normalization of constitutive materials of the machine as it does upon their "digital" and functional qualification. An iron mineral that had not been sufficiently laminated and deterritorialized would show unevenness from pounding that would falsify the ideal profiles of the key and the lock. The smoothing out of the material must remove the aspects of its excessive singularity and ensure that it behaves in a way that will take the molding of formal imprints exterior to it. We should add that this molding, in this sense comparable to photography, must not be too evanescent, and must keep a consistency that is its own and that is sufficient. There again we encounter a phenomenon of standard deviation, bringing into play both a material consistency and a theoretical diagrammatic consistency. A key made of lead or of gold might bend in a steel lock. A key brought to a liquid state or to a gaseous state immediately loses its pragmatic efficiency and falls outside the category of technical machine.

This phenomenon of formal threshold will recur at every level of intra- and extramachinic relations, particularly with the existence of spare parts. The components of technical machines are thus like the coins of a formal money, a similarity that has become even more manifest because computers have been used both to conceive and to execute such machines.

These machinic forms, this smoothing out of material, of standard deviation between the parts and of functional adjustments would tend to make us think that form takes precedence over consistency and material singularity, since the reproducibility of technological machines seems to require that each of its elements be inserted into a preestablished definition of a diagrammatic sort. Charles Sanders Pierce, who characterized the diagram as an "icon of relation" and attributed to it the algorithmic function, suggested an expanded vision that is still adaptable to the present perspective. Pierce's diagram is in effect conceptualized as an autopoietic machine, thus not only granting it a functional consistency and a material consistency, but also requiring it to deploy its various registers of alterity that remove what I call the machinic nexus from a closed identity based on simple structural relations. The subjectivity of the machine is set up in universes of virtuality that everywhere exceed its existential territoriality. Thus do we refuse to postulate a subjectivity intrinsic to diagrammatic semiotization, for example, a subjectivity "nestled" in signifying chains according to the famous Lacanian principle: "A signifier represents the subject for another signifier." There does not exist, for the various machine registers, a univocal subjectivity based on rupture, lack, and suture, but rather, ontologically heterogeneous modes of subjectivity, constellations of incorporeal universes of reference that take a position of a partial enunciator in domains of multiple alterity that it would be better to call domains of "alterification." We have already encountered certain of these registers of alterity:

the alterity of proximity among different machines and among parts of the same machine

the alterity of internal material consistency

the alterity of formal diagrammatic consistency
the alterity of evolutive phyla
the agonistic alterity among war machines, which we could expand to include the "autoagonistic" alterity of desiring machines that tend to their own collapse, their own abolition, and, in a more general way, the alterity of a machinic finitude

Another form of alterity has been taken up only very indirectly, one we could call the alterity of scale, or fractal alterity, which sets up a play of systematic correspondence among machines belonging to different levels.[4]

Even so, we are not establishing a universal table of forms of mechanical alterity because, in truth, their ontological modalities are infinite. Such forms are organized by constellations of reference universes that are incorporeal and whose combinatories and creativity are unlimited.

Archaic societies are better armed than white, male, capitalistic subjectivities to map this multivalence of alterity. In this regard I would refer the reader to the exposé by Marc Augé showing the heterogeneous registers to which the Legba fetish in the African Fon society refers. The Legba is set up transversally in

a dimension of destiny,
a universe of life principle,
an ancestral filiation,
a materialized good,
a sign of appropriation,
an entity of individuation, and
a fetish at the entrance to the village, another on the door of the house, and then at the entrance to the bedroom after initiation, and so forth.

The Legba is a handful of sand, a receptacle but at the same time the expression of the relation to others. It is found at the door, at the market, on the village square, at the crossroads. It can transmit messages, questions, answers. It is also the instrument of relation to the dead or to ancestors. It is at the same time an individual and a class of individuals, a proper name and a common name. "Its existence corresponds to the evidence of the fact that the social is not only a matter of relation, but a matter of being." Augé underscores the impossible transparency and translatability of symbolic systems. "The Legba apparatus . . . is constructed according to two axes. One seen from the outside on the inside, the other from identity to alterity."[5] Thus, being, identity, and relationship to the other are constructed, through fetishist practice, not only as symbolic, but also as ontologically open.

Contemporary machinic orderings, even more than the subjectivity of archaic societies, lack a univocal standard referent. But we are much less used to irreducible heterogeneity—or "heterogenicity"—of their referential components.

Capital, Energy, Information, the Signifier are so many categories that make us believe in the ontological homogeneity of referents—biological, ethnological, economic, phonological, scriptural, or musical referents, to mention only a few.

In the context of a reductionist modernity, it is up to us to rediscover that a specific constellation of reference universes corresponds to each emergence of a machinic crossroads, and that from that constellation a nonhuman enunciation is instituted. Biological machines advance the universes of the living, which differentiate themselves into vegetal becomings and animal becomings. Musical machines are founded on the basis of sonoric universes that have constantly been reworked since the great polyphonic mutation. Technical machines are founded at the crossroads of the most complex and the most heterogeneous enunciative components. Heidegger, who well understood that it was not only a means, came to consider technics as a mode of unveiling of the domain of truth. He took the example of a commercial airplane waiting on a runway: the visible object hides "what it is and the way in which it is." It does not unveil its "grounds" except "insofar that it is commissioned to assure the possibility of a transportation," and, to that end, "it must be commissionable, that is, ready to take off, and it must be so in all its construction."[6] This interpellation, this "commission," that reveals the real as a "ground" is essentially operated by man and is translated in terms of universal operation, travel, flying. But does this "ground" of the machine really reside in an "already there," in the guise of eternal truths, revealed to the being of man? Machines speak to machines before speaking to man, and the ontological domains that they reveal and secrete are, at each occurrence, singular and precarious.

Let us return to this example of a commercial airplane, no longer in a generic sense, but through the technologically dated model that was christened the Concorde. The ontological consistency of this object is essentially composite; it is at the crossroads, at the pathic point of constellation and agglomeration of universes, each of which has its own ontological consistency, marks of intensity, particular organization and coordination: its specific machines. The Concorde arises at the same time

- from a diagrammatic universe, with its theoretical "feasibility" plans;
- from technological universes that transpose this "feasibility" in terms of materials;
- from an industrial universe capable of producing it effectively;
- from a collective, imaginary universe corresponding to a desire sufficient to bring the project to term; and
- from political and economic universes allowing, among other things, the earmarking of funds for its production.

But the ensemble of these final, material, formal, and efficient causes, in the final analysis, don't make the grade! The object Concorde travels between Paris and New York, but it has remained bolted to the economic ground. This lack of economic consistency has definitively imperiled its global ontological consistency. The Concorde exists only within the limits of a reproducibility of twelve copies and at the root of the possibilist phylum of supersonics yet to come. That is already no small feat!

Why am I insisting so much on the impossibility of establishing solid grounds for a general translatability of various components of reference and for the partial enunciation of ordering? Why this lack of reverence toward the Lacanian conception of the signifier? It is precisely because this theorization, coming out of linguistic structuralism, does not get us out of structure, and prohibits us from entering the real world of the machine. The structuralist signifier is always synonymous with linear discursivity. From one symbol to another, the subjective effect emerges with no other ontological guarantee. As against that, heterogeneous machines, such as those envisioned in our schizoanalytic perspective, yield no standard being orchestrated by a universal temporalization. In order to illuminate this point I must establish distinctions among the different forms of semiological, semiotic, and encoding linearity:

1. encodings of the "natural" world, which operate in several spatial dimensions (those of crystallography, for example), and which do not imply extraction of autonomized encoding operators;
2. the relative linearity of biological encodings, for example, the double helix of DNA, which, based on four basic chemicals, develops equally in three dimensions;
3. the linearity of presignifying semiologies, which is developed in relatively autonomous parallel lines, even if phonological lines of spoken language always seem to overcode all the others;
4. the semiological linearity of the scriptural signifier, which imposes itself in a despotic manner upon all other modes of semiotization, which expropriates them and even tends to make them disappear in the framework of a communicational economy dominated by data processing (or, to be more precise, data processing at its current state of development, as this state of affairs is in no way definitive!); and
5. the superlinearity of asignifying substances of expression, where the signifier sheds its despotism, where informational lines can retrieve a certain parallelism and work in direct contact with referent universes that are in no way linear and that tend, moreover, to escape any logic of spatialized ensembles.

The signs of asignifying semiotic machines are "sign-points." Partly they are of a semiotic order, partly they intervene directly in a series of material machinic processes (for example, the code number of a credit card that makes a cash machine work).

Asignifying semiotic figures do not secrete only significations. They issue starting and stopping orders and, above all, they provoke the "setting into being" of ontological universes. An example may be found at present, in pentatonic musical ritornelli that, after a few notes, catalyze the Debussyan universe, with its multiple components:

the Wagnerian universe around Parsifal, which is linked to the existential territory constituted by Beyreuth;
the universe of Gregorian chant;
the universe of French music, with the rehabilitation of Rameau and Couperin for contemporary taste;
the universe of Chopin, thanks to a nationalist transposition (Ravel, for his part, having appropriated Lizst);
Javanese music that Debussy discovered at the 1889 World's Fair; and
the world of Manet and Mallarmé, which is linked to his stay at the Villa Medici.

And to these present and past influences should be added the prospective resonances constituted by the reinvention of polyphony since L'Ars Nova, its repercussions on the French musical phylum of Ravel, Duparc, and Messiaen, and on the sonic mutation unleashed by Stravinsky, its presence in the work of Proust, and so forth.

Clearly there exists no biunivocal correspondence between, on the one hand, signifying linear links or links of *arché-écriture*, according to authors, and, on the other hand, this machinic, multidisciplinary, multireferential catalyst. The symmetry of the scale, transversality, the pathic and nondiscursive character of their expansion, all these dimensions get us out of the logic of the excluded third term and comfort us by the ontological binarism that we had previously denounced. A machinic ordering, through its various components, tears away its consistency by crossing ontological thresholds, thresholds of nonlinear irreversibility, ontogenetic and philogenetic thresholds, thresholds of creative heterogenesis and autopoiesis.

It is the notion of scale that we should expand upon here in order to think fractal symmetries in terms of ontology. Substantial scales are traversed by fractal machines. They traverse them as they engender them. But it must be admitted that these existential orderings that they "invent" have already been there forever. How can we defend such a paradox? The reason is that everything becomes possible, including the recessive smoothing out of time described by René Thom, as soon as we allow for an escape from ordering outside of energy/space/time coordinates.

And there again, it falls to us to rediscover being's way of being—before, after, here and everywhere else, without however being identical to itself—of being eternal, of being processual, polyphonic, singularizable with textures that can become infinitely complex, at the whim of infinite speeds that animate its virtual compositions.

Ontological relativity sanctioned here is inseparable from an enunciative relativity. Knowledge of a universe in the astrophysical sense or in the axiological sense is possible only through the mediation of autopoietic machines. It is fitting that a foyer of self-belonging should exist somewhere so that whatever entity or whatever modality of being might be able to come into cognitive existence. Beyond this coupling of machine and universe, beings have only the pure status of virtual entities. The same goes for their enunciative coordinates. The biosphere and the mecanosphere, clinging to this planet, bring into focus a spatial, temporal, and energetic point of view. They make up an angle of constitution of our galaxy. Outside this particularized point of view, the rest of the universe exists—in the sense that we apprehend existence here below—only through the virtuality of the existence of other autopoietic machines at the heart of other biomecanospheres sprinkled about the cosmos. Even so, the relativity of spatial, temporal, and energetic points of view does not cause the real to dissolve into a dream. The category time dissolves in cosmological reflections about the big bang, while the category of irreversibility is affirmed. The residual object is the object that resists being swept away by the infinite variability of the points of view by which it can be perceived. Let us imagine an autopoietic object whose particles might be built on the basis of our galaxies. Or, in the opposite sense, a cognitivity constituting itself on the scale of quarks. Another panorama, another ontological consistency. The mecanosphere appropriates and actualizes configurations that exist among an infinity of others in fields of virtuality. Existential machines are on the same level as being in its intrinsic multiplicity. They are not mediated by transcendent signifiers subsumed by a univocal ontological foundation. They are themselves their own material of semiotic expression. Existence, insofar as it is a process of deterritorialization, is a specific intermachinic operation that is superimposed onto the advancement of singularized existential intensities. And, I repeat, there exists no generalized syntax of these deterritorializations. Existence is not dialectic. It is not representable. It is hardly even livable!

Desiring machines, which break with the great social and personal organic balances and turn commands upside down, play the game of the other upon encountering a politics of ego self-centering. For example, the partial drives and the polymorphously perverse investments of psychoanalysis do not constitute an exceptional and deviant race of machines. All machinic orderings contain within them, even if only in an embryonic state, enunciative nuclei [*foyers*] that are so many protomachines of desire. To circumscribe this point we must further enlarge our transmachinic bridge in order to understand the smoothing out of the ontological texture of machinic material and diagrammatic feedback as so many dimensions of intensification that get us beyond the linear causalities of capitalistic apprehension of machinic universes. We must also surpass logic based on the principle of the third excluded term and on sufficient reason. Through smoothing out, a being beyond comes into play, a being-for-the-other, which makes an existing being take consistency outside of its strict delimitation in the here and now. The machine is always synonymous with a constitutive threshold of existential territory against a background of incorporeal reference universes. The "mecanism" of this reversal of being consists in the fact that certain discursive segments of the machine begin to play a game that is no longer only functional or significational, but assumes an existentializing function of pure intensive repetition, what I have elsewhere called a ritornello function. Smoothing out is like an ontological ritornello and, thus, far from apprehending a univocal truth of Being through *techné*, as Heideggerian ontology would have it, it is a plurality of beings as machines that give themselves to us once we acquire the pathic and cartographic means of access to them. Manifestations not of Being, but of multitudes of ontological components are of the order as machines—without semiological mediation, without transcendent coding, directly, as "given-to-being"—as Donor [*Donnant*]. To accede to such a giving is already to participate in it ontologically, by rights [*de plein droit*]. This term of "right" does not crop up here by chance, so true is it that, at this proto-ontological level, it is already necessary to affirm a protoethical dimension. The play of intensity within the ontological constellation is, in a way, a choice of being not only for itself [*pour soi*], but for all the alterity of the cosmos and for the infinity of time.

If there must be choice and freedom at certain "superior" anthropological stages, it is because we shall also have to find them at the most elementary levels of machinic concatenation. But notions such as element and complexity are here susceptible to brutal reversal. The most differentiated and the most undifferentiated coexist amid the same chaos that, with infinite speed, plays its virtual registers one against the other, and one with the other. The machinic-technical world, at whose "terminal" today's humanity is constituting itself, is barricaded by horizons formed by a mathematical constant and by a limitation of the infinite speeds of chaos (speed of light, cosmological horizon of the big bang, Planck's distance and elementary quantum of action of quantum physics, the impossibility of crossing absolute zero). But this same world of semiotic constraint is doubled, tripled, infinitized by other worlds that, under certain conditions, ask only to bifurcate outside of their universes of virtuality and to engender new fields of the possible.

Desire machines, aesthetic creation machines, are constantly revising our cosmic frontiers. As such, they have a place of eminence in the orderings of subjectivation, which are themselves called upon to relay our old social machines that are unable to follow the efflorescence of machinic revolutions that are causing our time to burst apart at every point.

Translated by James Creech

Notes

1. Norbert Wiener, *Cybernetics; or, Control and Communication in the Animal and the Machine* (Cambridge, Mass.: Massachusetts Institute of Technology Press, 1948).

2. Francisco Varela, *Autonomie et connaissance* (Paris: Seuil, 1989).

3. Varela, *Autonomie et connaissance*; translation by James Creech.

4. Leibniz, in his concern to homogenize the infinitely large, and the infinitely small, thinks that the living machine, which he assimilates to a divine machine, continues to be a machine even in its smallest parts. This would not be the case for a machine made by human art. See, for example, Gilles Deleuze, *Le Pli* (Paris: Minuit, 1988); translated by Tom Conley as *The Fold* (Minneapolis: University of Minnesota Press, 1993).

5. Marc Augé, "Le Fétiche et son objet," in *L'Objet en psychanalyse*, ed. Maud Mannoni (Paris: Denoël, 1986).

6. Martin Heidegger, *Essais et conférences* (Paris: Gallimard, 1988), 1:9, 48.

Chapter 3

War, Law, Sovereignty—*Techné*

Jean-Luc Nancy

This chapter is my response to a request that came from the United States (from Tom Conley at the Institute for Research in the Humanities at the University of Wisconsin) for some reflections on "war and technology." To undertake this sort of reflection right in the middle of a war (as I begin writing, on February 26, 1991, the ground war in the Persian Gulf is on, and its future remains uncertain) may appear incongruous, even indecent. What counts today are, on the one hand, the immediate stakes—the dead, suffering of all kinds, and the great compassion that accompanies all wars (some of which, I hope, will be registered in these lines)—and, on the other hand, the political determinations—approbation, criticism, motives, and reasons that we can all perhaps still see as calling upon our responsibility.

But our responsibility is also already called upon in yet another sense: as the responsibility of thinking. Aside from the moral, political, and affective considerations, "war" as it is returning today is fundamentally, despite and because of its archaic character, a new reality. If you prefer, the return of "war," not as a reality of military operations, but as a figure (*War*) in our symbolic space, is a phenomenon of a singular and undeniable novelty, for this return is occurring in a world where the symbolism of war had appeared to be nearly effaced—something doubtless worth thinking about. And such thinking might well be a matter of some urgency: it is perhaps henceforth no longer a matter of knowing to what degree the war is a more or less necessary evil or a more or less adventurous good. It is a matter—and this for the *world*—of knowing to what symbolic space we can entrust what one calls freedom, humanity.

War, in Spite of Everything

Of course, symbolic War appeared to have been effaced only in that "world" formed by the group of nations constituting the planetary pole of "order," "law," and "development." The Third World has been continuously ravaged by armed conflicts, but it was as if these armed conflicts did not belong strictly speaking to the category of war, or as if the local character of these armed conflicts prevented them from attaining the full symbolic dignity of war. It seems that, ever since 1914, the notion of "War" has been applicable only to conflicts of "world" dimensions. I will come back to what this adjective implies. Let me first remark that this "worldliness" [*mondialité*] is determined less by the extension of the arenas of conflict (again, such conflict is strewn throughout the Third World) than by the world-encompassing role—economic, technological, and symbolic—of some of those States whose sovereignty is involved. War is necessarily the war of Sovereigns; there is no war without Lords of War. And this is what I wish to speak about here.

One might think that this is hardly the path of a questioning of "war and technology." We shall soon see, however, that in place of a concern with military technologies (about which there is nothing particular to think), the attention paid to the *sovereign* of and in war unveils the latter as the *techné*, the *art*, *execution*, or *operation* [*mise en oeuvre*] of Sovereignty itself. And Sovereignty is indeed a decisive, imperious, and exemplary punctuation of the entire symbolic of our Occident.

The war of States and State coalitions, "major war" ("true war"), does not begin in revolts against, or subversion of, States (or other collective entities), and does not manifest itself first of all as the wars or guerrilla actions of "peoples," but makes up by rights a part—and a privileged, exemplary part—of the exercise of the State and National sovereignties it presupposes. This is war properly speaking, such as it has been defined since the beginnings of our history (I will return to this). To the extent that such war can be simply distinguished from these other, more marginal, forms, we have thought we could circumscribe—if not suspend—it in the figure of the "cold" war and in nuclear deterrence.

But it is this war—or at least all of its symbolism—that is returning. Whatever name would be most appropriate to what is happening now, it has apparently been necessary to accompany, support, illustrate, and decorate these events with the signs, signifiers, and insignias of *war*. This return of "war" will have been irresistible, and not merely because of a negligent use of words.

For forty-five years—and to content myself with the figures most easily identified from a formal point of view—the wars of the Falklands and Grenada had prefigured such a return. (I owe to Robert Fraisse the decisive indication of this "return," and of the "wild contentment" with which, as he wrote, the Falklands war was accompanied.) The other armed operations concerned our ''world'' officially only under the sign of police interventions either in conflicts of the order of revolt, subversion, and ''civil war'' (whose name indicates, like the Greek *stasis* or the Roman *seditio*, that it is not a war between sovereigns, not war as such) or in confrontations between sovereignties we regarded as distant and often more or less dubious. (It would be necessary to expose in detail the claims, uses, manipulations, and aporias of sovereignty in the postcolonial world, as well as today in the post-Soviet world. And to link up with these a detailed consideration of *our* relations to all of this sovereignty, the concept of which is our own.)

But at present, there is *war*, and a ''world'' war in this new sense, that several of those Sovereigns are implicated whose titles we decipher in complex and contradictory ways. Even if what is at stake in this conflict is not merely North-South relations, the presence of these relations makes this world war even more world-ly, if one can put it this way. So there is war—for three months, the world will have had only this word in its mouth. But *what is there, precisely*, when there is war, and what is there *today*? This is what it is worth asking oneself.

The most surprising thing is not that there should be (*if there is*) this war. In any case, the most surprising thing is not that there should be this *combat* or this *battle*, whatever its genesis and modalities. The surprising thing is that the very idea of war should have regained among us a kind of right to conceptual citizenship. In other words, it is quite remarkable that the idea of legitimate State/National violence, having been so long suspected and even effectively unsettled by an at least tendential delegitimation, should have been able to regain (nearly) its full legitimacy—which means the legitimacy of Sovereignty, absolutely.

Some have said and written that, in terms of proper politicojuridical semantics, it is neither exact nor legitimate to use the word *war* with reference to the present case. I will return to this. But it is only very few people who have argued in this way, the juridical purists and the moral beautiful souls, while the general discourse, quite to the contrary, has flung itself with delight into the semantics, logic, and symbolism of war.

Of course, this semantics, logic, and symbolism had never been annulled. But once again, war seemed to remain withdrawn in the shadows into which the two previous ''world'' wars had plunged it. In distinction to previous centuries, the spirit of the times did not put the right to wage war on the highest level of the prerogatives of the State—as did, for example, up until the First World War, the use of the term *Powers* to refer to States.

Instead, the favor accorded to the idea of the ''constitutional State'' had directed our attention toward that which, in sovereignty, was considered exempt from violence and its explosive brilliance. Better, it directed our attention toward that place where the violence that would have presided over the institution of power was supposed to have been effaced, sublimated, or restrained. War seemed to repose in the peace of the feudalisms and nationalisms reputed to be defunct or obsolete. The brilliant glitter of sovereignty, too, was growing dull. Besides, we were finished with the ''ideologies'' of the ''withering away of the State'': the latter seemed to have entered into the age of *self-control*, declining by comparison with the worldwide complexes of technoeconomy, and confining itself to the—not terribly sovereign—role of regulatory administration, juridical and social.

As it happens, nationalism (and sometimes feudalism as well) is reviving on all sides. Its figures may be heroic or derisory, pathetic or arrogant, dignified or doubtful, but their vocations and destinies are always, in the final analysis, somewhat suspicious. To be sure, a worldwide recognition of the ''value'' of democracy, of democracy as norm, tends to regulate these affirmations of identity (and) of sovereignty. We imagine that the figures of State and Nation assert themselves not through violent, somber, and glorious gestures, but through spontaneous self-constitution in the bosom of a readily available general legitimacy.

It is well known, however—and this war, precisely, is reviving debate on this subject—that there is not (yet?) a supranational or prenational law. There is no ready-made ''democracy'' (that is, *foundation of law*) above nations and peoples. There is merely a law supposed to border the nation-states, less than certain about its foundations in universality, and indeed rather certain about its lack of sovereignty. In this law called ''international,'' it is in the *inter*, the *between*, that the problem resides, for the ''between'' is only graspable as a space empty of law, a space that is emptied of any kind of ''placing in common'' (and without which there is no law), but that is entirely structured by *both* technoeconomic networks *and* the surveillance of Sovereigns.

It is in this context that war has just made its figure known again in all its grandeur. Whether it is a ''war'' or a ''police action,'' whether or not it is even ''taking place'' as ''war'' in a certain sense does not matter very much. We have allowed, even ''needed'' (as has been said) a war. One step further, and we would have been making reference to the allegorical figures of Mars or Bellona, tempered for the needs of the occasion by a beautiful (i.e., arrogant) requirement of ''justice'' and ''morality.''

At last (I add this in taking up these lines again after the cease-fire), the victory parades have been announced, after the entire world adopted with delight the proud formula of the ''Mother of All Battles,'' which will have been the sovereign word of the defeated. In order for ''the logic of war'' (another sovereign expression) to be deployed, it was necessary that the possible return of this figure become perceptible, even if in a furtive (fleeting) manner. The States involved have skillfully exploited the virtualities emerging in the various manifestations of ''public opinion'': war was becoming necessary again, even desirable. The pacifisms were no longer anything but a matter of routine or accident, and moreover they were easily discredited for having recently failed to recognize the fascist peril, and for representing, in the final analysis, and since the beginning of the century, only the powerless obverse of the becoming-"world" of war.

But thus, just as pacifism is limited today to a *habitus* deprived of substance, the morality of which is articulated neither in terms of a law nor, above all, in terms of a politics (for its sole respectable dimension is pity; but the tragedy of war is not the only tragedy in this world—even if only war seems to have grandeur)—just so, on a very different register, the reaffirmation of war proceeds from a rediscovered *habitus* (a way of being, a disposition, manners, an *ethos*), played out anew in an altered context.

What sort of *ethos* is this? Of what is it composed? My initial response is simple: it is the very *ethos* of war, the disposition of manners, civilization, and thought that affirms war not solely as the means to a politics, but as an *end* consubstantial with the exercise of *sovereignty*, a sovereignty that in turn reserves for itself the exceptional *right* to wage war.

This response presupposes that we agree to call "police" the use of the force of the State with respect to its own laws, and "war" the exercise of a sovereign right to decide to attack another sovereign State. This convention is precisely the one that has just been reactivated, whether or not we have wanted to admit it (in terms of its Constitution, for example, France is not at war—and, moreover, who is at war in terms of what Constitution?).

A sovereign right has nothing superior to it (*superaneus*: which has nothing above itself). The right to war is the most sovereign of all rights, since it allows a Sovereign to decide that another Sovereign is his enemy, and to apply himself to the subjugation (i.e., the destruction) of the other, which means the removal of his sovereignty (his life into the bargain). It is the Sovereign's right to confront *ad mortem* his *alter ego*: this prerogative is not merely an effect of sovereignty but its supreme manifestation, and something of its very essence—as our entire tradition would have it.

Nothing else is valid, in the sovereign context of war, except certain conventions considered to contain war within a certain moral order (which was formerly a sacred order). But this order is not exactly superior to war: it is the very order of which war is a sovereign extremity, the lance head and the point of exception. (This is indeed why Rousseau, against nearly the entire tradition, did not want to see in war a special act of sovereignty, but "solely an application of the law"; Rousseau's sovereignty is in intimate debate with the exception and the explosive brilliance that cannot fail to haunt that sovereignty.)

Thus, war is itself susceptible to creating a new law, a new distribution of sovereignties. Such is indeed the origin of the majority of our State and National sovereignties and legitimacies. And such is also the point, from which revolutionary war inherited, by means of certain displacements, the essentials of the concept of the war of the State. (This begins with the wars of the French Revolution, a mixture of wars of the State and wars conducted in the name of a universal principle, against the enemies of the human race. From that time on, these wars raised the question of whether or not one could present a *universal* sovereignty.)

The right to war excepts itself from the sphere of law at the very point where it belongs to that sphere, like an origin and an end: in a point of *foundation*, insofar as we cannot think of foundation without *sovereignty*, nor think of sovereignty itself without thinking of it as exceptional and excessive. The right to war excepts itself from the sphere of law at the point where the sovereign fulgurates. Law does not possess this brilliance, but it needs its light as well as its foundational event. (This is why War is also the Event par excellence: not an event that could appear in a "history of events," that is, a list of the dates of wars, victories, and treaties [although, of course, such a list already tells us a lot about it], but the Event that suspends and reopens the course of history, the sovereign event. Our kings, marshals, and philosophers did not think otherwise.)

Yet this mode of inventing the sphere of law becomes inadmissible in a world that represents the law itself as its own "origin" or its own "foundation," whether in the name of a "natural law" of humanity or in the name of an irreversible sedimentation of the acquirements of a positive law that would have become bit by bit the law of all (whereas the soldiers of Year II of the Republic could still represent this foundation as a conquest to be made or to be remade). Hence our uneasiness at the idea of war, and in particular at the idea of the "just war," an expression that could at the same time submit war to law *and* law to war. (Nevertheless, and for the entire tradition [I will return to this], this expression is *in principle* redundant, as is *in fact* the expression "dirty war.")

Our uneasiness bears witness to the fact that our world—the world of the globalization of war—is displacing the concept of war, along with all the politico-juridical concepts of sovereignty. The "return" of war is occurring only in the midst of these displacements—and this is why certain people have attempted to say that it was not occurring at all. But our uneasiness bears witness also (and sometimes in the same people), I won't say out of either regret or nostalgia (although . . .), but out of a difficulty of doing without the sovereign instance, as far as its most terrible brilliance (for it is also the most brilliant). It is this resistance of sovereignty *within us* that I wish to examine—before attempting to comprehend toward what, toward what "other" of sovereignty we could move. As we shall see, this route passes by way of "technics."

I am not unaware of the precautions one must take in order to prevent this very simple program from ending up in oversimplification, or even in crudeness of thought. I will therefore take the precautions outlined below.

1. It is not my intention to reduce the history of the Gulf War to a mere sovereign decision in favor of war, whether it be a decision reached by one or several

of the actors involved. In a general context composed of endemic war, proliferating seditions, contested sovereignties, and multiple and conflictual policing forces (the rights and interests of states, minorities, and economic, religious, and international instances, and so on), a process was generated in which war and police action were intermingled, in which the one ceaselessly refers to the other. I do not claim to unravel to the end the part played by each of these terms, and this is probably impossible. Once again, everything is displaced, and the couple war/police action can no longer be handled simply—if ever that was possible. But I am interrogating in this couple what seems indeed to uphold obstinately, that is, ferociously, at the limit of the law itself, the requirement of war, which carries in itself and exposes the sovereign exception.

Now it is difficult to see how any mode of thinking at our disposal could render a satisfactory account of this logic of exception, this logic of the "sovereign" as what is "without law." The style of neo-Kantian humanism dominant today simply renews the infinite promise of a moralization of politics, even while it offers to the law the arms of a politics that remains to be moralized. The revolutionary style has foundered, stranding along with it all pretense to designate the subject of an *other* sphere of law and the emergence of an other history. As for the "decisionist" style, it is relegated to the heart of the "totalitarian" style. There is no way out, be it for thinking sovereignty *hic et nunc*, or be it for thinking beyond it. A history of the doctrines and problems of international law, of sovereignty and war since the first world conflict, would bear ample witness to this general difficulty.

For the moment, we can only draw the strict consequences of this balance sheet. I am not, then, interpreting the Gulf War in accordance with any of these schemas. I am positing only that there is an empty space that yawns between an always weak and unstable schema of the "war (police action) of law" and a reactivated (reheated?) schema of "sovereign war." And this space is not the space of a "people's war": the people, for the moment, are in the museums of the Revolution, or else in the museums of Folklore. This space is indeed a desert. It is not merely pitted with oil wells and bomb craters, it is the desert of our thought—the desert of "Europe" as well as the desert of the desolation and economic and cultural injustice that is becoming worse in the Persian Gulf and elsewhere, cutting across all laws and wars. In the end, it is true that *the desert is growing* (Nietzsche). For a long time I have hated the morose delectation taken by certain people in ruminating this sentence. But I admit it: the desert is growing, and so is the sterility of the dominant humanism, not even any longer militant, but arrogant with the arrogance of the weak, as it finally lets burst forth its own irresponsibility.

I do not lay claim here to the invention of another mode of thought: I wish only to situate its exigency, its extreme urgency. We are already in an *other* mode of thought, or rather it precedes us, and the war is showing us that we have to catch up with it.

2. If it is clear that my preference (which I hold in reserve here) was not in this case for war, I am nonetheless quite clearly conscious that the great majority of the partisans of the war thought of themselves as being partisans of a law superior or exterior to State sovereignties. In addition, many have borne witness to an exact meaning of the responsibilities of all the parties to the conflict. I am not taking any intentions to court here and I do not claim that all partisans of the war were simply dissimulating a preference for war beneath the cloak of law. Certain people did this, as is all too clear, and it is not particularly interesting. What is interesting is both *that* people have found themselves able to *affirm* war, and *how* they have been able to affirm it: in a manner more or less discrete or warmongering, cautious or complicated.

At the same time, however, it is not a question of suspending—by another turn of the tendency toward oversimplification in vogue today—the consideration of the interests and calculations that make up the economic stakes of the war, as much along an East-West axis as along a North-South axis. Moreover, denegation is in vain, everyone knows what's going on, and it is no longer necessary to be a Party member to share, willy nilly, some truths that come from Marx. It is not a matter of a simple "determination by the economy." Rather, whatever is the case as far as causalities are concerned, economy is in the process of exhausting perspectives, hopes, and *ends*. Whatever is not regulated in terms of economy belongs either to the sphere of a timid juridical protection (where it is no longer a matter of creating or founding a new legal order) or to the sphere of fantasmatic compensations (religion, sometimes art, and henceforth politics also). The return of the figure of War responds to an exasperated desire for legitimation and/or teleology in a situation where no one can believe any longer that the economy carries within itself its proper and universal legitimated *telos* (in this respect, what may remain of a difference between liberalism and state interventionism). In truth, at the moment when one believed oneself to be celebrating the "death" of Marx, his *political economy* (one could call it also his *economic war*) is nailing down our entire horizon. This political economy is not sovereign, but dominant, which is different. That is how politics commit suicide in a moralistic juridicism without sovereignty—or else, in order to serve domination more effectively, it attempts to regain its lost glory: and then we have War, ambiguous sovereign-slave of the economic "war."

I will come back to this world without *end*. Its critique must be no less *radical* than was that of Marx. But doubtless the path to this radicality no longer goes either by way of the invention of a new End or by way of the restoration of Sovereignty in general. To the contrary: such logic seems indeed to be the logic within which economic war uninterruptedly radiates sovereign War and vice versa.

3. It is true that in interpreting facts and discourses under the sign of the return of a dimension of war, the return of a bellicose posture or postulation that we had believed forgotten (if not repressed), I seem to ignore the restraint and prudence shown by those who have attempted to keep this war "limited." To be sure, there will have been few properly and directly bellicose discourses (and the language of "war, war, go away" could be heard on the side of the pacifist polemic). However, there will have been some remarkable effects in private discourse, and I will certainly not be the only one to have heard people boast of "the good it will do for the Westerners to rediscover their balls."

In taking up this text again after the cease-fire, I wish to add this: in view of the inequality of the participants in the battle, how can one not think that there was a need all along for the discourse of war, even if there was no desire for all of the aspects of war, but only for certain of its results? The "fourth army of the world" neither could nor wanted to fight. And the "first" fought above all in order to drown the very possibility of battle beneath a flood of bombs, at the price of compromising on its heroism by limiting its own losses. This prevents neither death and destruction nor, above all, the enormous disparity of their quantities in the one camp and in the other. But these quantities do not affect (at least not at first) the symbolic dimension of the war: the latter articulates itself solely in terms of *victory* and *defeat*, of sovereignty affirmed, conquered, or reconquered. (To that degree, this [un]certain war led to an [un]certain conclusion. At present, Iraq is minting coins with the inscription "Victory is ours," while the United States, Great Britain, and France are preparing military parades. Granted, all of this is pure facade. Moreover, the postwar period is propagating civil war, at least in Iraq and Kuwait, and reigniting the economic war. It remains the case, however, that "the facade" is not nothing in political constructions and the collectivity in general.)

It is indeed true that I presuppose the interpretation of a certain number of details, which would extend, for example, from the approval of the war by the national parliaments (a supererogatory measure in a police action, as also in a true state of emergency and great military threat) to all of those indices provided by the semantics, styles, and accents of so many discourses devoted to urgency, danger, sacrifice, national obligation, martial virility, the sublime of dark thunder, and the release of primitive forces (in a major French journal I read: "But how can one do battle effectively if one does not liberate within oneself one's primitive instincts?"; after all, this sentence is of course irreproachable, taken as it is and in the ordinary context of our culture, and even though it bears witness to the fact that this "ordinary" context is in a state that inclines toward vulgarity). One must join to these the discourses of the sacred mission: on each side, God was involved, *monotheos* contra *monotheos*—just as in the appeals to "foundation" of a new order or reign.

I do not bother to collect the public and private documentation, which is obviously massive. Interpretive violence is hardly necessary in order to decipher here the presence of a warlike symbolic and fantasmatic system, more or less discreetly mixed with reasons of law and policing. These reasons are not thereby disqualified, but the symbolic and fantasmatic system ought to be placed in its proper light.

In addition, one must not forget the role played on all sides by the political desire and need to make up for former lost wars (Vietnam and Sinai, although the two cases are quite different). In the case of the United States, the most powerful of the adversaries, the stain that had to be washed away was not merely the humiliation common to all defeats, but a war that had made war disgraceful.

Finally, one cannot forget the taste—so clearly displayed during the preparations for, and the first phase of, the war—for the spectacle of epic beauty and heroic virtue. After all, these images did not differ from all those of war movies. I do not want to add my voice nonetheless to the critics of the "society of the spectacle," who have not failed to characterize this war as "spectacular" (this denegation being symmetrical with the one operated by the discourse of law). The images of war are part of the war—and perhaps war is itself like a movie, before a movie could even come to imitate war. Before the horror and the pity, which the end requires, there would be no war without the warlike impetus of the imaginary. Its spectacle is inextricably linked with mechanical constraint, sometimes stupefied, that makes the soldier advance. The psychologists of the American army have taken pleasure in explaining (on television) that the "boys" were not marching for a cause, be it law or democracy, but solely in order not to back down in front of their companions. The resources of honor and glory are already by themselves of the order of the "spectacle," and one cannot undo them by means of a brief denunciation of a modern age of general and commercial simulation. (Moreover, and as always happens with this type of discourse, in reading certain criticisms of the "spectacle-war" one would have been justified to wonder what nostalgia was being broadcast for the good, true, grandiose war of former times.) What is at stake in the "spectacular" quality of the war is something much more fundamental, which extends all the way to the extremities of our entire culture (in which Islam takes part), and probably even beyond.

I am not claiming that the Epic is returning—neither the Homeric nor the Napoleonic variety, nor even the Epic that was associated, for example, with the combats of Rommel and Montgomery, of Leclerc or Guderian. (All the same, there has been some talk of the "legendary past" of the units and vessels that still bore in the Gulf the aura of their forebears from the First World War.) For a return of the Epic, much more would be required, but this "much" is not sufficient to prevent any affirmation and celebration of the war. There remain at least a number of facets of the explosive glittering of the sovereign. In war, we salute (for the instant, the time of a lightning flash) an explosive, incandescent, and fascinating sovereignty. But is not that an essential part of what we think we have lost in general: the explosive brilliance, the figures of the sun? Our world represents itself as lacking neither power nor intelligence nor even quite grace, but there is no doubt that the lack of Sovereignty structures an essential part of its representation and therefore of its desire.

Sovereign Ends

What has returned or what remains with the war is obviously not of the order of military technologies, for such technologies have been used all along, in all the quasi wars, guerrilla wars of liberation and their suppressions, and operations of political, economic, and juridical police. But above all, whatever is true of properly military technologies is that much more true of the technologies called "civilian" when they are employed for military ends. The difference between the two is almost impossible to establish. Psychology itself is a weapon and, reciprocally, military research has enabled countless developments in the field of civilian technology (for example, in the domain of sleeping pills). Perhaps a specific difference truly begins, on the one hand, at the level of purposes of mass destruction (but one suspects how much even this criterion could be difficult to handle, at least concerning material destruction, which can intervene in civilian activities), and on the other hand, and above all, at the level of symbolic markings (uniforms and the insignias of armies). There are uniforms outside of the army, but where they are found, there is also something of the army, as a more or less latent principle or model. Up to this line of demarcation, all the technologies involved, from the fabrication and use of a rifle or sword to the logistic and strategic management of entire armies, provide nothing that would allow us to specify the idea of *war* as such.

This is indeed why in a sense there is no specific question of the technologies of war. Here, as in any technological domain, there are only technological questions, which never enable us to question or to think "technology." In the first days of this war, the placing in the limelight of certain featured technologies made it possible for us to observe how discourses that were favorable or unfavorable to technology had nothing to do with technology, but rather corresponded to the positions taken on the problems and the aporias of the war itself. (The English word *technology*, as opposed to the French word *technique*, serves well to suggest that there is a logic proper to the technical, to which the hurried activity of discourses regulated by "sense" and "values" is almost never geared.) We have celebrated the strength of technological fire, of the electronic, chemical, and mechanical complex that produces, among other things, the "missile," the latest arrival in the immemorial series of martial emblems: sword, helmet, cannon, and so on. We have congratulated ourselves on the possible self-limitations of this same power, in a discourse of "surgical" war that responded to the thesis of law: to proclaim that one is holding oneself within limits more restrictive than those of international conventions is to render more credible the interpretation in terms of "police." We have deplored the terrible possibilities offered by new technologies (for example, "vitrification" by air-bombs; as for cluster bombs of phosphorus and napalm, they are already well known). Finally, we have feared the recourse to technologies banished by the Geneva accords: the NBC group, outlawed today, but the effective use of which in the past and the catastrophic virtualities of which play an obvious strategic role. In this respect, moreover, the accords on the means of war continually manifest the fragility of the law that underlies them: not only is it infinitely difficult to legitimate, in the name of humanitarian principles, the distinction between some arms and others, but the collision—that is, the contradiction—between these principles and those of war does not cease to be perceptible, and to refer the "right to war" back to its foundation in the sovereign exception. It is clear that it is not from the law—or from the logic—proper to war that one can infer, for example, the prohibition of extending destruction to genetic heritages. But this is also why one has been able to observe, up to a certain point in this war, the discreet progress made by the idea of a tactical use of nuclear arms in response to the chemical threat of Iraq. (It might well be that atomic weapons were, more than was generally admitted, one of the major stakes in this conflict: such weapons, their possession, and their use in the next war.) One could develop considerations parallel to these concerning the protection of civilians. But then, all of this is *well known*, which means of course that *one wants to know nothing about it*.

Thus, there is no "question of technology" proper to war, any more than there is a "question of technology" in general, no doubt, as long as one understands by that a question posed to technology, a question that would treat technology as a subject, applying to it criteria that do not pertain to its own domain. War conducted with missiles is neither better nor worse than war conducted with ballistas: it is in each case a matter of *war*. And communication is neither better nor worse when it is carried by optic fibers than when it is carried by messengers on foot. Of importance is that we find out what something like "communication" means. If "technological" civilization displaces the concepts of war and communication (and health, and life, and so on) such as they are, then it ought to be a question of these concepts themselves, of their "becoming-technological" within a general space of the becoming-technological of the world. But it is not a matter of evaluating new instruments for the unchanged purposes of a world that is forever old.

War may well be a privileged terrain for shedding light on the inanity of all considerations on technology that do not proceed from this requirement (one must admit that such considerations are indeed the most numerous). It is clear that no given technologies are directly responsible for the war, any more than the war is responsible for technologies that are not properly its own—even if technologies provide the means to war and war stimulates technological progress. The ethical, juridical, and cultural problems posed by civilian technologies (nuclear or biological technologies, for example) are no less acute than those posed by certain types of weaponry. It is even probable that the disparity between these two levels of the problem has become substantially reduced since, for example, the epoch of the invention of artillery (and this reduction doubtless attests to the "becoming-technological" of the world for which we have to account). But the interminable discourses of celebration or execration of technology, founded on "values" that are obstinately foreign to this becoming, can in the final analysis only mask what is at stake in "war," as well as what is at stake henceforth in, for instance, "medicine" or "the family."

One has not correctly raised the "question of technology" as long as one still considers technologies as means in the service of ends. All the "questions," except for the problems of the technicians as such, are posed thus with respect to ends: practical or ethical, political, aesthetic, and so on. As long as war is itself considered as a means in the service of ends (political, economic, juridical, religious), it falls within such a logic. And this is what is truly at stake in Clausewitz's formula that "war is the pursuit of politics by other *means*." This formula signals a modern mutation of the thought of war. This mutation takes its distance from—and no doubt denegates in a more or less confused way—the "classical" thought of war, for which war is the exercise, operation [*mise en oeuvre*], and extreme expression of sovereignty. As I have already said, this thought is always, in principle, the only rigorous thought of war. The displacement undertaken by Clausewitz must still be completed: it is perhaps the *end* of war.

From the traditional Western perspective that we still generally adopt, war is the technology par excellence of Sovereignty: its operation and its supreme execution (*end*). A "technology" in this sense is not a means, but a mode of execution, manifestation, and effectuation in general. It is precisely that mode of performance that distinguishes itself from the "natural" mode as its double and its rival in perfection. As soon as one has recourse, in a contemporary usage that goes back to Heidegger (and more discreetly to Nietzsche, if not to the German Romantics), to the Greek terms of *phusis* and *techné*, it is in order to give specific names to these "modes of perfection," by distinguishing them both from "nature" as an ensemble of materials and forces, governed by its own laws, and from "technology" as an "artificial" means to arrive at ends. *Phusis* and *techné*—one could say "blossoming" [*éclosion*] and "art"—are two modes of performance, and are in this respect the *same* (not identical) in their difference: the same of performance in general, of operation or execution. Consequently, one is confronted twice here with the "same" of the *end*, not with two modes of purposiveness but with two finishings (like a "hand-stitched" finishing and a "machine-stitched" finishing—a comparison that fits well to recall further the hierarchy we set up "quite naturally" between the two modes of finishing). Further, these two modes continually refer to each other in a double relation called, since Plato and Aristotle, *mimesis*: the one does not ''copy'' the other (''to copy'' is precisely impossible in this case), but rather each replays the play of the end(s), as the art or blossoming of finishing.

Finishing consists in executing—*ex-sequor*, to pursue until the end—by carrying something out to the extremity of its own logic and its own good, that is, to the extremity of its own being. From the perspective of Western thought, being in general, or rather being proper or being properly speaking [*l'être propre ou l'être en propre*] in each of its singular effectuations or existences, has its substance, its end, and its truth, in the *finishing* of its being. This trait belongs indeed so obviously, for us, to ''being'' in general (or to reality, to effectivity) that it seems strange to insist on stating this redundancy.

Being is not being halfway, we think; it is being present, achieved, complete, and finished, each time as final, terminal, executed for itself. The entire problem, if there is a problem, is to know if the execution, the finishing, is finite or infinite, and in what sense of these terms. As we shall see, the questions of technology and war point in the final instance to this problematic articulation.

Phusis and *techné* are thus the being of being, the same playing itself out twice, in a difference to which we shall have to return. For the moment, let us add simply that *history* is the general system of a twisting or displacement that affects this difference.

If, then, there is a ''question of technology,'' it begins only at the moment when technology is accounted for as finishing of being, and not as a means to some other end (e.g., science, mastery, happiness), and consequently as an end in itself, *sui generis*. Technology is a ''purposiveness without purpose'' (i.e., without extrinsic purpose) of a kind that it remains perhaps for us to discover. And it is to such a discovery that our history exposes us, as the becoming-technological of being or its finishing.

What falsifies in principle so many considerations on technology is that they seek its principles and its ends outside of it—for example, in a ''nature'' that, however, itself continually enters into a becoming-technological. Just as we continue to relate ''nature'' to some sovereign Power—the creation and glory of a Power named God, Atom, Life, Chance, or Humanity—so we continue to obtain from technology, and for technology, a *Deus ex machina*, that other sovereign Power that, nonetheless, the most habitual tendencies of our representations lead us to designate as a *Diabolus ex machina* (this is the entire story of Faust). *Ex machina*, the *Deus* becomes *diabolicus* because it is no longer the ''technician of nature'' or the Natural Technician, that is, the one who relates all things to an End or to a Finishing that is absolute, transcendant or transcendental, and sovereign.

Perhaps Leibniz was closer to a first clear consciousness of technology in seeking to bring to light a *machina ex Deo*—unless it is more fitting to combine him with Spinoza, in a *Deus sive natura sive machina*, after which the ''death of

God'' signifies the rigorous execution of the program thus sketched, the *machina ex machina* (*ex natura*), the one that *does not finish finishing*, the law and the implications of which it remains for us to think through.

One must consider here, further, the singularly marked position that our thought assigns to war between ''nature'' and ''technology.'' (Note also the ''quite natural'' ambiguity of our comprehension of such a sentence: Is it a matter of *war* considered as occupying an intermediary position *between* nature and technology, or is it instead a matter of a war that *takes place between* nature and technology? Precisely, we are ready to think these two things together.) War is that which is most and/or least ''natural.'' It has its origin in the ''most brutal instincts'' and/or in the coolest calculations, and so on. This position is not without relations with the position we give to ''beauty'' between ''art'' and ''nature.'' This position, at once problematic and privileged—a position itself replayed twice between two orders, art and war, that are considered to be somehow opposed the one to the other—is not indifferent, and we will return to it below.

All considerations of ends lead us back to sovereignty. The power of ends, as the power of the ultimate or the extreme, resides in a sovereignty. And every end, as such, is ordered by a sovereign end (a ''sovereign good''). For all our thought, the End is in Sovereignty, and Sovereignty is in the End. The absolute transcendence, abyss, or mystery of supreme ends in the entire tradition—for example, the impossibility of determining the ''content'' of the Platonic Good or the Kantian Law—is caught exactly in this circle: that which is sovereign is final, that which is final is sovereign.

Sovereignty is the power of execution or of finishing as such, absolutely and without any further subordination to something else (i.e., to another end). Divine creation and the decision of the Prince compose its double image: to make or unmake a world, to impose one's will, to designate an enemy. This is indeed why, when there is war, the executive power attains to a state of exception (emergency), foreseen by the legislative and controlled by the judicial branches, but arriving in spite of everything, *in principle as in fact*, at the extremity of decision and power (powerful decision and decisive power) where it accomplishes most properly its ''executive'' essence, the sovereign essence of being, whose ''power'' it constitutes (as Prince, State, Nation, People, Fatherland).

This is why it is necessarily somewhat less than rigorous to accuse a sovereign power of *wanting* war, in the sense of having a *will* to war. The execution of this will to war does not merely constitute one of the proper ends of the executive organ. It represents also the extreme mode of its ends, and thus no longer an organ, but sovereignty itself in its finishing—at least as long as we think sovereignty in accordance with the only concept of it that is at our disposal. In war, something goes immediately beyond all possible goals of war, be they defensive or offensive, to the performative effectuation of the sovereign as such in a relation of absolute opposition with another Sovereign. War is indissociably the *phusis* and the *techné* of sovereignty. Its law—the exception of its law—has indeed as its pendant the law of grace, but with the latter, the sovereign neither identifies itself nor executes itself vis-à-vis the *other*.

If necessary, we could find confirmations in the particular symbolic or fantasmatic charges of various instruments and machines of war. And it is difficult to deny that, if the Gulf War provided an occasion for an explicit discourse of sovereignty only in an inhibited and fleeting—that is, denegational—manner, nonetheless it aroused an exceptional deployment of images of tanks, jets, missiles, and helmeted soldiers, images saturated with a symbolic charge, and even *images of symbolic saturation* itself, a symbolic saturation that may well constitute a trait of sovereign finishing.

Objects lose their symbolic character to the degree that their technicity grows, at least the technicity that one understands in terms of functionality (i.e., as a means); but this functionality does not prevent the object from being symbolically (or fantasmatically) invested anew. This is the case with a sickle, a hammer, a set of gears, and even a circuit board. But nowhere else today (or, for that matter, yesterday) can the symbolism adhere to the function in such an immediately obvious manner as in the image of the arms of war. And without a doubt, this adherence comes about because the image of arms does not present a tool of destruction but first of all the affirmation of the sovereign right of the sovereign power to execute a sovereign destruction, or to execute itself in destruction, *as* Destruction (of the other sovereign). It is hardly a function at all, but a destiny: to give and to receive collective death, death sublimated into the destiny of the community, the community identified in a sovereign exposition to death. (Could Death be the true Sovereign in this entire affair? We will come back to this.)

It is thus that war borders upon art—not upon the art of war (i.e., upon the technology of the strategist), but upon art taken absolutely in its modern sense (i.e., upon *techné* as a mode of execution of being, as its mode of finishing in the explosive brilliance of the beautiful and the sublime, double and rival in sovereignty of the blossoming of *phusis*). (Moreover, *phusis* no longer takes place except as mediated by *techné*, if indeed one can say that it ever took place "in itself," otherwise than as the image of the sovereignty of *techné*.) Doubtless, the aestheticization of the spectacle of war comes also from denegation and dissimulation. But this manipulation does not exhaust an aesthetic (a sensible presentation) of the destiny of community: the death of the individuals allows them to be preserved in the figure of the Sovereign—Leader or Nation—where the community is finished. War is the monument, the festival, the somber, pure sign of the community in its sovereignty.

War is in essence collective, and the collectivity endowed as such with sovereignty (the Kingdom, State, or Empire) is by definition endowed with the right to war (''*bellum particulare non proprie dicitur*,'' Thomas Aquinas, S.T. IIaIIae, 123, V). The entire history of the concept of war would reveal a constant play of its determination in a double relation with *res publica* (the commonwealth as a *good* and as an *end* in itself) and with the *Princeps* (the principle and principate of the sovereign authority). Not only is the latter in charge of the former, the prince in possession of the armed force necessary to the preservation of the republic, but also the commonwealth as such is supposed to present and represent his absolute and final character, his sovereignty, and its armed force is supposed to carry the flag of his glory.

It is at this precise point that the law of the republic—of any kind of republic down to our own times—inevitably touches upon the exceptional status of the prince, whatever the form of government. To the present day, democracy has not profoundly displaced this schema: it has only suppressed or repressed it into the shadows of its own uncertainties (that is, of its uncertainty as to its own sovereignty, an uncertainty that until today has remained part of its very substance). Thus, the schema of the sovereign exception does not cease to return, like the repressed, and like the perversion of democracy, whether it be in the innumerable *coups d'État*, large or small, that inhabit its history, or in its becoming-totalitarian (the exception thus transforming itself into a doubling of the structure of the State by another, which incarnates true sovereignty).

Since the First World War, however, it is democracy as such—such as it ended up presenting itself as the general principle of humanity, if not utterly as humanity's End—that has been thought to be endowed with the right to war, thereby transforming war into the defense of the *res publica* of humanity. This presupposed that a neutral country (the United States—when one thinks about it, neutrality is a strange form of sovereignty, as long as there are several Sovereigns) decides to depart from its neutrality in the name of the rights of man, and that it designate explicitly as its enemy not a people or a nation, but governments judged to be dangerous to the good of all peoples (''civilized'' peoples). In world war, democracy does not go to war against a Sovereign (Germany, or the countries of the Alliance), but against bad leaders.

(Note added April 6, 1991: Today, in the face of the suppression of the Kurds by the same sovereign leader on whom was inflicted the ''police action'' of law, the Powers are hesitating between respect for his sovereignty within his borders and affirmation by the ''international community'' of a right of interference in the matters of particular countries. I can think of no better way of illustrating the inconsistency and the aporias, today, between the notions of ''international law'' and ''sovereignty.'' This said, it is obviously not these conceptual difficulties that motivate the diverse judgments and hesitations. Still, these difficulties express the real state of a world that is at a loss in the face of the notion of sovereignty, but that nonetheless does not know how to displace or go beyond this notion.)

In order for the decision to go to war—against Germany in the past, against Iraq today—in the name of the rights of man to become a decision (and not a mere wish), it was necessary that this decision take form and force in and by the sovereignty of a State—and/or an alliance of States. When one or more States speak in the name of the rights of man, and make use in this name of the prerogatives of the *jus belli*, this is still and always the fact of a sovereign decision (or of an alliance of such decisions). In a sense, it is even a surfeit of sovereignty in comparison with the Prince of the tradition. This is indeed why, in the Gulf War, the coming and going between the authority of the United Nations and the authority of the United States (and, if one takes this seriously, the authority of some other States as well) has been so complex and so simple at the same time, so delicate and so indelicate. The legitimacy without sovereignty of "international law" needed a sovereign *techné*—and *not*, as one was led to believe, a mere means of execution. But this Sovereign needed in turn the legitimacy of the rights of man in order to ground the decisions and claims that, like the principles and promises of the law, were to be of world-encompassing dimensions. (Of course, despite being called upon by the Sovereign power, the law remains nonetheless without foundation, i.e., without sovereignty and without "finishing.")

Here as everywhere else it is solely a question of the Public Good and of Peace. To this too, the entire history of our thought of war testifies, from Plato and Aristotle to the Christian and then Republican doctrines. When Henry Kissinger declared not long ago that "the goal of all wars is to ensure a durable peace," his judgment was supported (or weakened?) by the authority of twenty-five centuries of philosophical, theological, ethical, and juridical repetition. War in the West always has peace as its *end*, to the point where it is necessary even to "battle peacefully," according to the expression of Augustine to Boniface (*epist.* 189). The state that gives itself war as the end of its structure and formation is Sparta, which Plato submits to a severe critique. No doubt, the principle of ultimate peace has been inflected more than once, not merely (as is obvious) in fact, but in theory itself (for example, by mixing with the logic of "peace" a logic of religious conversion or a logic of the occupation of territories claimed as one's patrimony). Nonetheless, the general theoretical regulation of war in the West remains that of pacificatory war (the motif of which has been extended to cover the exportation of certain colonialist forms of "peace"). Western war denies itself as sovereign *end*, and this denial, as usual, constitutes its avowal.

It would be necessary here to take the time to analyze the complex play of the three great monotheisms of "the Book"—which are also the three monotheisms of "community" and thus of Sovereignty. Israelite monotheism (at least until the destruction of the Temple) and Islamic monotheism, each of them in its own complex ways, reserve a place for a principle of war that does not confuse itself with the peace of the peoples. Christian monotheism proposes a different complexity, which in particular mixes up the model of the *pax romana* with the model of the war against the Infidels. Even as a religion of love, it does not confuse itself simply with a principle of peace, for there are the enemies of love and, above all, divine love has an essence different from that of human love. In its becoming-modern (which set in, to be sure, at the very beginning of its development), Christian love, taking up into itself the irenic principle of Greek philosophy (which presupposes the breakdown of the Epic and which mimics this breakdown in the installation of the logos), becomes entirely a principle of peace and of peace in the universal rights of people. The god of love loses its divinity here bit by bit, and love in peace loses its sovereignty. The peace of humanism is without force or grandeur; it is but the enervation of war.

Deprived of the Temple and of a site of sovereignty, of State and soil, Jewish monotheism was supposed, precisely for this lack of sovereignty, to be annihilated in a beyond of war itself. Inversely, resuscitated in the service of Western Statism and Nationalism, Islamic *djihad* reanimates a flame of Crusade in the face of peace and of the police of law. At every moment, however, Jewish monotheism can identify and assign anew the Sublime Sovereignty it puts into play—and at every moment, Islam can ruin in contemplation and self-abandonment the Absolute Sovereignty that seals its community. Thus, triple monotheism is installed in the double regime of, on the one hand, the war of Sovereignties and, on the other hand, in each of these, the tension between its execution and its retreat.

But the symbiosis of triple monotheism and its other/same, philosophical monologism, presents itself thus under the sign of a war of principles: sovereign war (the war of the three gods against the triple god) versus pacificatory war—or, again, the confrontation between sovereign war and sovereign peace. This confrontation is present in philosophy itself, between an absolute request of peace (demanded by the *logos*) and an incessant recourse to the schema of *polemos* (demanded also by the *logos*, which mediates itself through this schema). But the sovereignty of peace remains a promised and/or ideal sovereignty, while the sovereignty of war is already given. It leaves behind in the *polemos* the trace of divine refulgence, of epic song, and of royal privilege. Thus, still today, in philosophy as in all the nerves of our culture, war for peace cannot stop being war for war and against peace, whatever course it may take. Technology in the service of peace cannot not be taken up again into the *techné* of sovereignty—that is, into Sovereignty as *techné*, execution and finishing off of community (reputed not to be a community through *phusis*, that is, to have its "nature" in *techné politiké*; in this regard, it would be necessary of course to show how, with the Greeks, *techné politiké* splits itself in principle into sovereign *techné* and *techné* of justice or law, thereby constituting the program of their impossible suture).

It follows that when one lays claim to it, the sovereignty of peace is not in a symmetrical relation to the sovereignty of war. Peace would be, instead, the "supreme" good the supremacy of which could not manifest itself as such, either in glory, power, or collective identification. There remains the pale dove. Peace would be the supremacy of the supreme absence of distinction, the absence of exception at the heart of a rule everywhere indefinitely and equally closed upon itself. But thus peace cannot fail to carry, for our entire culture, a connotation of *renunciation*. For finally, Sovereignty properly so called requires the incandescence of the exception and the identifiable distinction of its finish. (And, in fact, will one have ever seen an identifiable peace, presentable in person, except under the name and insignias of an empire—*pax romana, pax americana*?)

Thus, the sovereignty of law, which was supposed to structure peace, is inevitably, and in however small a measure, sovereignty by default, whereas true sovereignty takes place not merely in plenitude but in excess and *as* excess. This fundamental disposition prevents war, still today, from ever becoming simply a technique for putting force in the service of right, without being always also the *techné* of sovereign affirmation.

It is therefore not sufficient to keep returning indefatigably to the *ultimate* aim of peace, any more than it is sufficient to denounce the illusion of such an aim and to align oneself with the realism of power. These two faces of the same attitude have essentially regulated our relationship to the recent war, by means of the total or partial repression of what I have attempted to recall. To content oneself with this situation is to prepare the wars of the future—and without even wanting to know if the restraint that has been shown, in certain respects at least, in the conduct of this war (to the point where there wasn't "truly" a war—although there has been all the desirable destruction) does not represent a still modest step toward the complete "relegitimation" of war, a "relegitimation" whose conditions of possibility would not be so far off as some would have liked to believe, or to lead others to believe.

I will be told that the accent placed thus on the symbolic order of sovereignty denies simultaneously or by turns the authenticity of the needs of law and the play of economic forces. Not at all, as we shall see. Rather, a symbolic order that is so widely and deeply woven into an entire culture produces, as is well known, all its effects in the real (and thus, for example, in economy and law—but in truth, none of these "orders" comes simply from the symbolic *or* from the real). And it is important not to misconstrue these effects. Just as much as art, war properly so called is absolutely archaic in its symbolic character, which means doubtless that it escapes from being merely a matter of "history" in the sense of the progress of a linear and/or cumulative time. But it returns when it is a matter of opening anew in this time a certain space: the space of the presentation of Sovereignty. This "archaism" (once again, like that of art) thus obeys laws that go deeper within our civilization than those of a vexatious survival. But it is precisely because it is inconsistent to treat war like an annoying holdover from a bygone age, always tendentially effaced in the progress and project of a world-encompassing humanity, that it is all the more important and urgent to think what is at stake in its "archaism," and to think this for ourselves today.

(Nevertheless, the examination of this space of sovereignty and war obviously requires things quite different from what I have just sketched out. There is an enormous program of work here, in particular on the register of analyses of the "sacred." Sovereignty has always been mixed up with the "sacred" through the energy of exception and excess, but the implications of sovereignty have not yet been as clearly thought out as have been the implications of the sacred itself [as if under the effect of an obscure interest in not knowing too much about the sovereignty that is always at work]. There is much work to be done here also on the register of a psychoanalysis that would manage, in a way different from what psychoanalysis in general has attempted thus far, to treat collectivity or community as such [which Freud seems indeed always to submit, *volens nolens*, to the schema of the Sovereign], not to speak of the sexual differentiation unfailingly operated by "manly" war.)

Ecotechnics

All of that having been said, it remains that the belated persistence or the reinvention of war does not produce itself outside of history—even if our epoch is the epoch of a great suspension of the historicity on whose stream we were hitherto borne along. The conflict between the police and the *bellum proprie dictum* is also the effect of a historical displacement that is of great importance and of great consequence for war.

The first of the "world" wars corresponds to the emergence of the schema of an order of world proportions that imposes itself on the Sovereigns themselves. The police war is delocalized; it has less to do, for example, with the borders of the sovereign States than with the multiple forms of the "presence" of these States all across the world (interests, zones of influence, and so on). In this way, the police war becomes also a confrontation of "worldviews": a worldview is never the attribute of Sovereignty; by definition, the latter is above all "view," and the "world" is the imprint of its decision. The Powers had the world as the space given for the play of their sovereignties. But once this space is saturated, and the game closed off, the *world* as such becomes the theme of a problem. It is no longer certain that the *finishing* of this world can be envisaged as was that of the world of the Sovereigns. The world—that is, man, or again *world* humanity—is neither the sum of humanity nor the installation of a new sovereignty (contrary to what humanism sought and desired to the point of exhaustion). The police war of world humanity puts in question directly the *ends* of "man," whereas sovereign war as such exposed the end. And just as war and art were the *technai* of sovereignty, so world humanity has no *techné* of its own:

however thoroughly "technological" our culture may be, it is only *techné* in suspension. It is not surprising that war haunts us.

(The invention of world war can be seen to be not only the corollary of the development of a world market, but the result of the wars that accompanied the engendering of the contemporary world: on the one hand, the two American wars—the War of Independence and the Civil War—as wars in the tradition of sovereign war, bearing the self-affirmation of a new and distinct Sovereignty [the wars and/or national foundations of the nineteenth century, and first of all of the new Germany, partake of this model, later bequeathed to the colonies] and, on the other hand, the war of liberation in the name of humankind, its "natural" rights and fraternity, war as the French Revolution invented it. This second model no longer corresponds exactly to the sovereign schema: it oscillates between a general revolt against the very order of Sovereigns [named *tyrants*, a term that makes an appeal, in the ethicojuridical tradition, to a possible legitimacy of rebellion] and a policial administration of humankind, which itself suppresses the abuses of its governing instances.)

The world-encompassing state of war expresses then—as its cause or effect—a simple need: the need for an authority that goes beyond that of sovereigns endowed with the right to wage war. Strictly speaking, this need has no means of being received into the space and logic of sovereignty. More precisely, it can be analyzed in two ways: either (1) this authority corresponds to a world Sovereignty, which in this case cannot be in a state of war with anyone on earth (but only with all the Galactic Empires of science fiction, which shows that indeed there has thus far been only one model from which to extrapolate) or (2) this authority is of another nature, and of another origin (and of another *end*) than that of sovereignty.

From the League of Nations to the United Nations, there has been ongoing discussion of the aporias of such a "suprasovereignty"—both from the point of view of its legitimate foundation and from the point of view of its capacity to endow itself with an effective force. To different degrees, analogical problems are posed for the transnational organizations of the states of Africa and Asia. In still another way, Europe is encountering the problem of a sovereignty that would be *inter-*, *trans-*, or *supra*national—a problem that in this case does not in principle concern war except for the transformations, already in progress, of the two great blocs of military alliances.

It should by now be apparent that the problem is radical. It is not a matter solely of combining the requirements of coordination—that is, of international cooperation—with the respect of the sovereign rights of states. Nor is it solely a matter of inventing new politicojuridical forms (whether one goes in the direction of the deliberative Assembly or in the direction of a world Federation, one will not have overcome these aporias). The forms adhere to the contents and their grounds—and it is, moreover, one of the tasks of law and its formalism to bring to light the work of grounding in the construction of concepts. But precisely, *law does not have a form for what ought to be its proper sovereignty.*

The problem is well posed, in an exact and decisive manner, at the very site of Sovereignty—or of the End. The problem is not that of managing sovereignty: in essence, sovereignty is untreatable. But above all, the untreatable essence of sovereignty in fact no longer belongs to a world that has become worldly in the sense adumbrated above. The problem is thus indeed that of grounding in an entirely new way something that has neither reasons (why? for what? for whom is there or ought there be a *worldly* world?) nor an appropriate model. Worldly man—man according to humanism—is man exposed to a limit or an abyss of grounding, end, and exemplarity.

However relative, however mixed up it may be with the turns and returns of so many particularist claims to an unconvincing "sovereignty," the "return of war" expresses essentially a need for or drive to sovereignty. Not only do we have nothing other than models of sovereignty, but Sovereignty in itself is one of the principal models or schemas through which the becoming-worldly of our "civilization" is occurring. It is the model or the schema of "that which has nothing above itself," of that which cannot be gone beyond, of the unconditional or the nonsubordinable, of all this *quo magis non dici potest* where origin, principle, end, finishing, the leader, and explosive brilliance meet. But the man of the "world" is another sort of extremity, another *quo magis* . . . , to which this model, in truth, no longer corresponds.

Only law seems to escape from the domination of the model thus maintained by default. Law installs itself in fact from the very first between first principles and final ends (the sovereign space is the space of the figure; the juridical space is the space of the interval). Law consents to subordinate principles and ends to an authority other than its own, and this consent belongs to its structure. It escapes from the model, then, only in order to designate anew the places of the model: at the two extremities of principles and ends. Sovereignty cannot stop haunting us, since at these extremities law situates of itself the instance of the exception and the excess, which is also the instance of exemplarity.

In the domain of exemplarity, the exception always serves or gives the rule. (Thus, the model of the Sovereign Warrior was not simply an invitation to battle. Still, the entire history of sovereignties is in the end a history of desolations.) But after the dissolution of exemplarity, one is left with, on the one hand, the exception into which the rule is reabsorbed and, on the other hand, a rule without example (law), that is, an unfinished rule.

We have failed to recognize what was truly at stake in war because, ever since the invention of the "natural" law theory of the rights of man (an expression in which "natural" really means "technological"), we have insisted both on closing our eyes to the absence of foundation of the law and, symmetrically, on mistaking the foundational role of Sovereignty as the schema of the exception (divine creation, originary violence, founding hero, royal race, imperial glory, martial sacrifice, spirit of the work, subject of one's own law, subject without faith or law). Beneath the judgment that war is (sometimes "necessary") "evil," we have repressed the truth that war is the model of the *techné* that executes and finishes off, insofar as the End is conceived as a sovereign end. In a symmetrical way, beneath the judgment that law is a formal "good" without force, we have repressed the truth that, if it is not governed by a Sovereignty, law without model or foundation represents a *techné* without end, a thing with which our thought cannot deal (except to ghettoize it, for good or ill, in "art") and of which it is *afraid*. We will not have answered the question of war, except by means of ever more war, until we have traversed this problematic field.

How is one to think without end, without finishing, without sovereignty—and without nonetheless resigning oneself to a weak, instrumental, servilely humanist thought of the law (and/or "communication," "justice," the "individual," the "community," all of which concepts will remain feeble until we have found some answer to this question)?

However, it is not sufficient to ask the question in this way. It is well known that the "world (dis)order," if it is without reason, end, or figure, nonetheless has all the effectivity of what one calls "planetary technology" and "world economy": the double sign of a single complex of the reciprocity of causes and effects, the circularity of ends and means. The without-end, indeed, but the without-end in millions of dollars and yen, in millions of thermies, kilowatts, optical fibers, kilooctets. If the world is a world today, it is a world first of all under this double sign. Let us call it here *ecotechnics*.

It is remarkable that the country that hitherto has been the symbol of triumphant ecotechnics concentrates also within itself the two figures of the sovereign State (supported by the arch-law of its foundation as by the hegemony of its domination) and law (present in the foundation, and thought of as structuring "civil society"). The Soviet world that was its counterpart was supposed to represent the revolution that reverses and goes beyond this triple determination, restoring a social-human *whole* to itself as to its end. In fact, this world was not that of the State, nor that of law or of ecotechnics, but a painfully distorted imitation of the three and their interrelationships, in the service of the pure appropriation of power. But it is not less remarkable that these two entities shared, in their difference and in their opposition (in the "cold war" of two Sovereigns differently frozen or fixed), a kind of asymptote or point of tendential convergence, something that one would have to call *sovereignty without sovereignty*, to the extent that this word and its schema remain inevitable: the supreme domination of what would have neither the incandescence of the origin nor the glory of accomplishment in a sovereign presence; neither God, nor hero, nor genius, and yet the logic of the subject of exception, subject without law of his own law, and an execution, an indefinite and interminable finishing off of this logic. Ecotechnics could be the last figure without figure of this slow drifting of the world into sovereignty without sovereignty, into finishing without finishing (off).

And thus, the recent war would have been simultaneously a powerful resurgence of sovereignty (while warning us to expect others perhaps) *and* the opening up of a passage, from the very interior of war, to the regime (or reign?) of sovereignty without sovereignty. Just as there has been an attempt to give the war the form of a police action, there has been an attempt to turn the conceptual couple, victory and defeat, into a matter of negotiation, where what is at stake is ''international law'' as the guarantor of ecotechnics. At the same time, all sides have refused to count the dead clearly (to say nothing of the distinction between dead soldiers and dead civilians): given the plausible report of at least one dead (in the North, in the West) for five hundred dead (in the South, in the East), it seems that victory and defeat are both becoming as insupportable as they are insignificant. And finally, the true regime of this war—*as everyone knows perfectly well*—is revealing itself to be the ecotechnical war, the destructive and appropriative confrontational maneuvering that has no sovereign incandescence—but that does not for all that lag behind real war as far as power and the technologies of ruination and conquest are concerned.

The *class struggle* was supposed to be the other of both sovereign war and ecotechnical war. It is frequently claimed that this struggle is no longer taking place, or no longer has a place in which to take place: this claim amounts to the declaration that there is no conflict outside of sovereign war (whose very return one denies) and ecotechnical war (which one calls ''competition''). Thus, war is found nowhere, but the laceration, ruination, police violence, and savagery that are caricatures of ancient sacred violence are found everywhere. War nowhere and everywhere, related neither to a supreme end outside of itself nor to itself as supreme end. Ecotechnics too is indeed in one sense pure *techné*, the pure *techné* of nonsovereignty; but because the empty place of sovereignty remains occupied, encumbered by this void itself, ecotechnics does not accede to another thought of the end without end. For sovereignty, ecotechnics substitutes administered ruination under the control of ''competition.''

Thus, ecotechnics is henceforth the name of ''political economy,'' for we cannot conceive of *politics* without *sovereignty*. There is no longer any *polis* when *oikos* is everywhere: world housekeeping, as of a single household—''Humanity'' as Mother, ''Law'' as Father.

But it is well known that this great family has neither Father nor Mother and is in the final analysis no more an *oikos* than a *polis*. (Ecology: What semantics, what space, what world can it offer?) What it is can be summarized in three points:

1. It implies a triple division that is by no means a sharing [*partage*] of sovereignties: the division of the rich from the poor, the division of the integrated from the excluded, and the division of North from South. These three dimensions do not coincide as simply as one sometimes suggests, but this is not the place to speak of that. I wish to underline only that they imply struggles and conflicts of great violence, where all considerations of sovereignty are in vain and borrowed. But if the "class struggle" itself also conceals its schema (and without a doubt it is no longer admissible, at least in a certain historical dimension), then nothing remains to prevent violence from being camouflaged as ecotechnical competition. Or rather, only naked justice remains: but what is a *justice* that would be neither the *telos* of a history nor the endowment of a sovereignty? It is necessary then to learn how to think it through in this empty place.

2. Ecotechnics breaches, weakens, and disorders all sovereignties—except for those that in reality coincide with ecotechnical power. The nationalisms, whether they be of ancient lineage or of recent extraction, deliver themselves up to the painful mimicry of a mummified Sovereignty. The current space of sovereignty—which obviously no cosmopolitanism can recuperate (for cosmopolitanism was always the dreamy inverse of the sovereign order) and which is also the space of the finishing off of *identity* in general—is merely a distended space full of holes, where nothing can come into presence any more.

3. Ecotechnics privileges—with more or less hypocrisy or denegation, according to the case, but not simply without pertinence—a primacy of the combinatory over the discriminating, the contractual over the hierarchical, the network over the organism, more generally the spatial over the historical, and within the spatial a multiple and delocalized spatiality over a unitary and concentrated spatiality. These motifs compose an epochal necessity (the effects of fashion here are secondary, and do not invalidate this necessity at all). Today, thought passes by way of these motifs, if it is thought *of this world*, that is, precisely, of this worldly world without sovereignty. But this is indeed why the entire difficulty of this thought is concentrated here. One could give it the following general formulation: How not to confuse this *spacing* of the world with either the *exposure* of significations or a *gaping* of sense?

In effect, either significations are exposed and diluted to the point of insignificance in the ideologies of consensus, dialogue, communication, or values (where sovereignty is thought to be nothing but a useless memory) or a surgery without sutures holds open the gaping wound of sense, in the style of a nihilism or aestheticizing minimalism (where the gaping wound itself emits a black glow of lost sovereignty)—and this is not less ideological. Neither justice nor identity is put back in business.

In order to think through the *spacing* of the world (of ecotechnics), one has to face head on, without reservations, the end of sovereignty, instead of making believe one has evacuated or sublimated sovereignty. This spacing of the world *is itself the empty place* of sovereignty. That is, the empty place of the End, the empty place of the Common Good, and the empty place of the Common as a good. If you like, again, in a word: *the empty place of Justice* (at the foundation of the law). When the place of sovereignty is empty, neither the essence of the "good" nor the essence of the "common" nor the common essence of the good can be assigned any more. Moreover, no essence can be assigned any more, no finishing off: only existences are finite, which is also what the spacing of the world means.

How to think without a sovereign End? This is the challenge of ecotechnics—a challenge that up to now has not been taken on, and that this war is finally beginning perhaps to make into an absolute urgency. To initiate a response, it is necessary to begin again with this: ecotechnics deluges or dissolves sovereignty (or rather the latter implodes in the former). The point is to endure its empty place *as such*, and not to expect a return or substitution. There will be no more sovereignty: this is what *history* today means. The war *with* ecotechnics lets us see the henceforth empty place of sovereign Sense.

This is besides why ecotechnics itself can summon in this empty place the figure of Sovereignty. Thus, the gaping of the foundation of law, and all the questions in suspension around the exception and the excess, can be forgotten in the sovereign refulgence borrowed for the duration of a war by the power properly without law that polices the world order and watches over the price of primary resources. Or else, on another register, the empty place of the one who recites an Epic is occupied by the sovereign figure of a prophet of the Moral Law (which can at the same time transform itself into the teller of small familiar epics, of the kind, "our boys from Texas"). Facing this figure, another figure attempted to reanimate the Epic—with the sole aim of taking part in the ecotechnical power of the masters of the world. On the one side as on the other, it was necessary that the models—the identifiable examples—of the sovereign allure come to guarantee the best presumption of Justice or of the People.

The empty place of Sovereignty will continue to give rise to (more or less successful) substitutions of this type until this place *as such* is submitted to questioning and to deconstruction—that is to say, until we have asked *without reserve* the question of the end, the question of the extremity of finishing off and of identity, which is henceforth the question of a *nonsovereign sense* as *the very sense of the humanity of humans and the worldliness of the world.*

Doubtless, the relation between nonsovereign sense and the archaism of Sovereignty, a relation we *ought* to invent, is still more complex than this. The very spacing of the world, the opening of a discontinuous, polymorphous, dispersed (i.e., dislocated) spatiotemporality, presents to Sovereignty something of this Sovereignty itself: on the near side of its figures and their imperious and avid presences, it has always also, and perhaps from the very first, exposed itself as spacing, as amplitude (of an explosion), as elevation (of a power), as distancing (of an example), as the place (of an appearing). And this is why the same motifs can serve by turns the ardent and nostalgic recall of sovereign figures—War before all others—and this access, which we ought to invent, to the spaciousness of spacing, to the (dis)locality of the location. (For example, and at the price of an excessively brief summary: the same process appeals by turns to America and to Arabia, exposing diversely intermingled fragments of reality, none of which is simply "Arab" or "American," and which compose a wandering, strange "worldliness.")

The worldly world of ecotechnics itself also definitively proposes, even if in obscurity and ambivalence, the thoroughgoing execution of sovereignty. And *thoroughgoing* means here going to the extreme end of its logic and movement. Until our own times (but this could continue), the extreme end always finished itself off in War, in one way or another. But it is henceforth apparent—*this* is our history—that the extreme of sovereignty situates itself *further* still, and that the commotion of the world means that it is not possible not to go further. War itself, supposing that one can detach from it the appropriation of wealth and power, *does not go any further than the explosive brilliance of death and destruction* (and in the final analysis, the voracious appropriation of war is perhaps not so extrinsic as it appears to the sovereign work of death). (Or else, if it is necessary to go further in the same logic, prolonging war beyond war itself, and death beyond death, there is the night and fog of extermination.) Death, or identification in a figure of death (that is, the complete movement of what we call *sacrifice*, of which war is a supreme form), is the aim of sovereignty, which always finishes by appropriating it to itself.

But thus, it has not gone far enough. Being-exposed-*to*-death, if it is indeed the "human condition" (finite existence), is not a "Being-*for*-death" as destiny, decision, and supreme finishing off. The finishing (off) of finite existence is a nonfinishing [*in-finition*], which overflows on all sides the death that contains it. The in-finite sense of finite existence implies exposition without explosive brilliance, discreet, reserved, discontinuous, and spacious in this, that the existent *does not attain to the sovereign extremity*.

"*Sovereignty is NOTHING*": Bataille exhausted himself in trying to say that—which doubtless one cannot but exhaust oneself in (not) saying. What this sentence "means" no doubt cuts off one's breath (at any rate, I do not wish to get into it further here), but whatever it "means," it certainly does not mean that sovereignty is death—to the contrary.

Here, I will say merely this: the sovereign extremity signifies that there is nothing to "attain to," neither "accomplished performance" nor "achieved completion," nor "finishing"—or rather, that *for a finite finishing, the execution is without end*. The worldly world is also the finite world, the world of finitude. Finitude is spacing. Spacing "executes" itself infinitely. Not that this recommences endlessly—but that *sense* is no longer in a totalization and presentation (of a finite or accomplished infinite). Sense is in this: not to have done with sense.

In this "nothing" there is neither repression nor sublimation of the violent burst of sovereignty: there is, *not to be done with*, an explosion and a violence from beyond war, the lightning of *peace*. (I owe to Jean-Christophe Bailly the suggestion to render pacific the eagles of war.)

In a sense, it is *technology* itself. What one calls "technology," or again what I am calling here ecotechnics (henceforth to be liberated in itself from capital), is the *techné* of "finitude" or of spacing. No longer the technical means to an End, but *techné* itself as in-finite end, *techné* as the existence of the finite existent, its brilliance, and its violence. We are talking about "technology" itself, but about a technology that of itself raises the necessity of appropriating its sense *against* the appropriative logic of capital, and *against* the sovereign logic of war.

Finally, the point is not that war is "bad." War is "bad"—absolutely so—when the space where it unfolds no longer permits the glorious and powerful presentation of its figure (as figure of the death of all figures), when this space constitutes spacing, intersection of singularities, and not the confrontation of visages or masks.

It is here that our history encounters today its greatest danger and its greatest chance. Here, in the still badly perceived imperative of a world that is in the process of creating its *global* conditions to render untenable and catastrophic the sharing [*partage*] of riches and poverty, of integration and exclusion, of all the Norths and all the Souths. Because this world is the world of spacing out, not of finishing; because it is the world of the intersection of singularities, not of the identification of figures (of individuals or of masses); because it is the world in which, in short, sovereignty is exhausting itself (and at the same time resisting with gestures at once terrifying and derisory)—for all of these reasons, and from the very heart of the appropriative power of capital (which itself occasioned the decline of sovereignty), ecotechnics obscurely points to the *techné* of a world in which sovereignty is NOTHING. A world in which spacing out can coincide neither with display nor with laceration but merely with "intersection."

This is not given as a destiny: it is offered as a history. Ecotechnics is still to be *liberated*, as *techné*, from "technology," "economy," and "sovereignty." At least we are beginning to learn, however slowly, what is after all the lesson of war, law, and "technological civilization": that the index, the theme, and the motive of this liberation are contained in this (provisional) statement: that *sovereignty is nothing*. And that, consequently, the multiplicity of "peoples" need not be engulfed either in the hegemony of one Sole people or in the agitation of the desire of sovereign distinction for all. In this way, what has up to now remained unthinkable could become thinkable: a political articulation of the world that escapes from these two dangers (and for which of course no model of "Federation" would be serviceable). In this way, law could expose itself to the nothing of its own foundations.

Thus, it is a matter of going all the way to the *unexampled* extremity of the "nothing" of sovereignty. How to think, how to act, how to do without a model? This is the question that is avoided and yet posed by the entire tradition of sovereignty. One must take seriously the sense in which execution without model or end is perhaps indeed the essence of *techné* as a *revolutionary* essence, once "revolution" too has been exposed to the *nothing* of sovereignty.

What if each *people* (this would be the revolutionary word), each *singular intersection* (this would be the ecotechnical word) substituted for the logic of the sovereign (and always sacrificial) model not the invention or the multiplication of models—upon which wars would immediately follow—but a completely different logic, where singularity was at once absolutely valid and not susceptible to exemplification? Where every *one* would be "one" only in not being identifiable in a figure, but in-finitely distinct through spacing *and* in-finitely substitutable through the intersection that doubles spacing. That could be called, to parody Hegel, the *World Singularity*. It would have the right without right to say the law of the world. *Peace* comes at this price: the price of sovereignty abandoned, the price of what goes beyond war, instead of remaining always within it.

I am well aware of it: this cannot be conceived. It is not for us, not for our warmongering thought modeled on the sovereign model. But there is decidedly nothing on the horizon but an unheard-of, inconceivable task—or war. All forms of thought that still want to conceive of an "order," a "world," a "communication," a "peace," are absolutely naive—when they are not simply hypocritical. To appropriate one's own time has always been unheard of. But everyone sees clearly that it is time: the disaster of sovereignty is henceforth sufficiently extended and sufficiently common to destroy our illusions.

April 1991

Postscript, May 1991

In the midst of the general climate of the "humanitarian aid" installed by the perverse game being played by the protagonists of the war, "sovereignty" is more present than ever. (Does Saddam have the right to it? Who grants it to him? What is he doing with it? And that of the Kurds? And the Turks? And what is a border? A police force? Or else, a bit further away, nuclear capacity as a sovereign matter of Algeria, or the accord between the USSR and the eight republics to regulate, despite everything, the tense play of their sovereignties; or else, Kuwait returned to its sovereignty *both* for the savage settling of accounts *and* for the shameless recruitment of Philippino and Egyptian manpower; or else, what is the sovereignty of Bangladesh where a cyclone just made five million people homeless? And so on.) The proliferation of these ambiguities—which are indeed those of the *end* of sovereignty—makes me afraid of being misunderstood when I say that we ought to go (or that we already are) beyond its model and its order. I do not for a moment mean thereby to demand that a Kurd, an Algerian, a Georgian, or, for that matter, an American should abandon the identity and independence of which these proper names function as signs. But what cannot fail to cause a problem is the question of what *signs*, precisely, are concerned here. If Sovereignty has exhausted its sense, and if it appears everywhere as doubtful, tricky—empty—then it is necessary to rethink from the beginning the nature and function of such a sign. For example: What is a *people*? The Iraqi "people," the Corsican "people," the Chicano "people," the Zulu "people," the Serbian "people," the Japanese "people": Is the same concept operative in all of these cases? If there is a "concept," does it imply "sovereignty"? And what about the "people" of Harlem, or the "people" of shantytowns in Mexico, or the peoples or populations of India or China? What is an "ethnic group"? What is a religious community? Are the Shiites a people? And the Hebrews and/or Israelis and/or Jews? And the "ex-East Germans"? What are the relations between a "sovereign" people and a "popular" people? Where to place tribes, clans, brotherhoods? And—I insist on this—social classes, levels, margins, milieus, networks? The quasi-monstrous multiplication of these questions is the mark of the problem of which I am speaking. The sovereign model and the instance of the law do not address this problem, they only deny it. But it is a matter of *worldliness* as a proliferation of "identity" without end and without model—and perhaps it is even a matter of *"technology" as techné of a new horizon of unheard-of identities*.

Translated by Jeffrey S. Librett

Chapter 4

Our Narcotic Modernity

Avital Ronell

Time Release

Our question concerns, once again, technology. It is a locational question. Where does the technological take place? Somewhere off the map, like airwaves, on cellular formations or in the deterritorialized zones of Arabia. One is tempted to say that technology strives to escape its detection systems, outsmarting itself according to new trajectories of the invisible. What this suggests, in more philosphical tones, is that the dream of exteriority has to be rethought. Some years ago, Martin Heidegger discovered that we latecomers were addicted to technology. Without consistently demonizing its hold, he asked under what conditions we could arrive at a free relation to technology. He posed the *Gestell* about which so much has since been written. Here, again, we encounter a problem of positionality. Where did he locate *Gestell*? Was the human *Dasein* invaded by its effects or was *Gestell* somehow a disposable limit, external and surmountable? Or, on the contrary, has *Dasein* become *dependent* upon *Gestell*? In *Being and Time*, Heidegger suggests the beginnings of such a thesis when he momentarily suspends the elaboration of care in order to treat the questions of addiction, urge, willing, and wishing.[1] Heidegger never names the substance of addiction but rather shows that addiction is addicted. This poses a problem that nonetheless opens a crucial passage through which *Dasein* must go. The problem is multifaceted. While the urging on of the urge belongs to vitality, and thus to a more positive form of addicted addiction, less passive or debilitated in nature, it is still inauthentic. This is due in part to the fact that addiction is content with what is merely available, ready-to-hand, and it never surpasses this limit; addicted, *Dasein* goes nowhere fast. Being on the run, it dumps understanding along the way, the way one flushes evidence when there's a bust. Though cast in a somewhat different idiom, Heidegger essentially plots this movement of an internalized *Gestell*. The restitution from an external frame to internal structuring, however, is something that never became an explicit philosophical theme for Martin Heidegger.

In a sense, freedom depends upon *Dasein*'s openness to anxiety, which addiction and urge are seen to divert. Addiction produces a parallel track to anxiety that closes it down. Again, Heidegger does not name the form of dependency or the object of urge to which *Dasein* can succumb. Nonetheless, this topos reemerges throughout his work with the relentless surprise of a textual symptom. In this essay, I try to locate these questions in our culture, implicitly relying on Heidegger's recircuiting of technology through addiction. If drugs have anything to do with such a technology, then we need to review their essence.

Drugs have been the focus of what Americans call "the other war." During the war in the Persian Gulf, a test site for high technologies, drugs were largely unaccounted for. This is because, in war, they shed their value as drugs and become another piece of equipment. A fighter pilot who might pop pills does so with a view to superior performance, technologizing himself into the war machine. It would be difficult to dissociate drugs from a history of modern warfare and genocide. One could begin perhaps in the neighborhood of ethnocide, with the American Indian, where the instruments of extinction were alcohol or viral infection. Drugs empirically participate in the analysis of warfare: methedrine, or methyl-amphetamine, synthesized in Germany, had a determining effect in Hitler's *Blitzkrieg*; heroin comes from *heroisch*, and Göring never went anywhere without his supply. It may be the case that there has never been a war that was not on drugs.

Yet what is it about drug addiction in particular that attracts the arrogance of war, or appears to compromise a technically calibrated culture? To shift the scale of questioning slightly, if only to achieve focus, What values is an electronic culture bound to protect? A technique of shifting is not extraneous to the topos at hand, but is rather commanded by it. Electronic programs—of which this inquiry is an effect, in view of its internal velocities, systems of presentation, and transmission of information—depend upon discontinuous sequencings for their intelligibility. By *electronic culture* I mean a certain cybernetic swipe at metaphysics. Of course, to a large extent, cybernetics has been superseded by the more sophisticated discipline of artificial intelligence, but it has had the lasting effect of retaining an essential distinction between human and machine. Yet before all man-machinic hybridizations, a technology of the human has already been in place. The focus here will be on the chemical prosthesis.

It is less a question of the toolness or instrumentality of mediatic incursion than of the relation to a hallucinated exteriority that these reflect, or rather the place where the distinction between interiority and exteriority is radically suspended, and where this phantasmic opposition is opened up. For this reason it is possibly less compelling to read the machine as an object than to observe the excription of metaphysical cravings to which it calls attention.

If the literature of electronic culture can be located in the works of Philip K. Dick or William Gibson, in the imaginings of a cyberpunk projection or a reserve of virtual reality, then it is probable that electronic culture shares a crucial project with drug culture. This project should be understood in Jean-Luc Nancy's and Blanchot's sense of *désoeuvrement*—a project without an end or program, an unworking that nonetheless occurs, and whose contours we can begin to read.[2]

In his book on possible limits of dying and getting high, Ernst Jünger, who turned Heidegger on to technology, writes about the drug drive.[3] While he begins to ask how the prosthetic subject is constituted, his thought is not all that remote from what Benjamin wrote about hashish, or even what de Quincey wrote about opium. It sometimes resembles Marguerite Duras's alcoholizations: it is a saturated text, pushing beyond the materiality of the book though not into any ideality. Drugs, for which Jünger effectively writes a manifesto, are the site of an allotechnology: technology's intimate other, sharing the same project of historical *désoeuvrement*. What Jünger says about the right to drugs, as well as about the supplementary interiority that they produce, has been explored elsewhere.[4] His book tries to design the conditions for a thanatorium of fractal dimensions of self: the experience of another limit, a fissure in the law of finality. Discovering a relation to death that might be other than skull and crossbones, Jünger wants to hitch a ride with Faust, whose first stop is at the kitchen of the witch. We are reminded by Jünger that the witches' brew is not merely the narcotica with which the expiring are usually let off, stupefied and always already out of it. Rather, the witches' brew offers Faust a powerful stimulant, opening an altogether other hallucinogenre *in* life, at the edge of being. The mysterious substance does not produce a virile meeting with nothingness, but creates an angle of exteriority in being.

Like Faust, who leaves the space of contained knowledge and control-room epistemologies, the probing scholar has to unhook in part from institutional apparatuses in order truly to consider these edges. The link between the electronic and drug cultures is compelling in part because drugs constitute a place of nonknowledge that has attracted the crudest interventions, and also because there has never been a war on drugs that is not carried by another type of drug (religion, patriotism, oil, TV). Where one can study the question of technological addiction via the positive technologies, including media and the machine, it is perhaps timely now to raise questions about the structure of addiction as such.

Toward a Narcoanalysis

When he wanted to formulate the task of a philosophy yet to come, Friedrich Nietzsche committed this thought to writing: "Who will ever relate the whole history of narcotica?—It is almost the history of 'culture', of our so-called high culture."[5] Our work settles with this Nietzschean "almost"—the place where *narcotics* articulates a quiver between history and ontology.

Addiction will be our question: a certain type of "being-on-drugs." This has everything to do with the bad conscience of our era. Baudelaire assimilates intoxication to a concept of work? Indeed, the plant puts you to work on a whole mnemonic apparatus. Intoxication names a method of mental labor that is responsible for making phantoms appear. In any case, it was a manner of treating the phantom, either by making it emerge or vanish. It was by working on Edgar Allen Poe that Baudelaire recognized the logic of the tomb to which he attached the stomach. The stomach *became* the tomb. At one point Baudelaire seems to ask: Whom are you drowning in alcohol?[6] This logic called for a resurrectionist memory, the supreme lucidity of intoxication, which arises when you have something in you that must be killed. Hence the ambivalent structure stimulant/ tranquilizer.

When the body seems destined to experimentation, things are no longer introjected but trashed: dejected. The body proper regains its corruptible, organic status. Exposed to this mutability, the body cannot preserve its identity, but has a chance of seeing this fall, or ejection, sublimated or revalorized. Nautilus versus the addict. When some bodies introduce drugs as a response to the call of addiction, every body is on the line: tampering and engineering, rebuilding and demolition, self-medication and vitamins become the occupation of every singularity. Sometimes the state has a hand in it. As Heidegger puts it, "Addiction and urge are possibilities rooted in the thrownness of Dasein."[7]

Crisis in immanence: drugs, it turns out, are not so much about seeking an exterior, transcendental dimension—a fourth or fifth dimension—as they are about exploring *fractal interiorities*. This was already hinted at by Burroughs's "algebra of need."

We do not know how to renounce anything, Freud once observed. This type of relation to the object indicates an inability to mourn. The addict is a nonrenouncer par excellence (one thinks of the way Goethe mastered renunciation), yet addiction does establish a partial separation from an invading presence.

The communication systems with the question concerning addiction are on, though each time beaming different signals, along edges of new line drivers. Much like the paradigms installed by the discovery of endorphins, being-on-drugs indicates that a structure is already in place, prior to the production of that materiality we call drugs, including virtual reality or cyberprojections. Our problem remains how to present a logic of something that is already there without resorting to the ontic. One of the implicit questions to emerge in this probe unavoidably concerns technology and, in Heidegger's terms, *Gestell*. What is *Gestell* in relation to the addictive hankering of *Dasein*? Would it not require remodeling in light of *Dasein*'s revision according to what in English is colloquially if oddly translated as "hankering" and "addiction"? Heidegger does make it clear that there is something like *Dasein*'s internalization of *Gestell* according to the chemical prosthesis. It appears that he thought about addiction (in *Being and Time*, in *Schelling's Treatise on the Essence of Human Freedom*), but not about the specificity of the technology of the drug. Yet, *Dasein* as addiction has everything to do with *Gestell*. Where in his work on *Gestell* Heidegger indicates man's blind dependency on technology in this phase of metaphysical disclosure, in his fundamental ontology there is a reading of "hooked pulsion" (*der Drang*) and dependency (*Nachhängen*). In pure pulsion anxiety has not yet become free, he argues, while in dependency anxiety is still bound. The question arises as an existential-ontological fundamental phenomenon that, Heidegger admonishes, is hardly simple in its structure ("in seiner Struktur nicht einfach ist"). The larger issues of will, urge, and craving are addressed where Heidegger zooms in on a regional mapping of *Dasein*'s dependency. It is at this place that Heidegger puts his rhetoric on tranquilizers, when concern, in its average everydayness, becomes blind to its possibilities "and tranquilizes itself with that which is merely actual." However, "this tranquilizing does not rule out a high degree of diligence in one's concerns, but arouses it."[8] The tranquilizer of *Sorge* acts as a stimulant, if only eventually to argue that Dasein closes off essential possibilities when it is shown to be addicted. Heidegger's rhetoric of drugs, as it were, is injected into the very place where *Being and Time* treats the problem of dependency. It turns out that when *Dasein* "sinks into addiction," there is not merely an addiction present at hand, "but the entire structure of care has been modified." This is nothing less than to locate the threat that addiction poses to Being.[9]

To gain access to the question of "being-on-drugs," we shall go the way of literature. I have chosen a work that exemplarily treats the persecutory object of an addiction. It does so within a fictional space, according to the fanatical exigency of realism. Like few other works of fiction, it brings out evidence of a pharmacodependency with which literature has always been secretly associated—as sedative, as cure, as escape conduit or euphorizing substance, as mimetic poisoning. There are many reasons for pressing literature on the narcotic question, but these are not essential: we could have just as easily followed the trajectory of *Rausch*, the ecstasy of intoxication, through the works of Kant, Nietzsche, and Heidegger on aesthetics. We could have traced the vertigo of the subject guided by the philosophemes of forgetfulness. Perhaps we would have arrived at the same rcsults. Still, it is the case that the singular staging of the imaginary—"literature" in the widest sense—has a tradition of uncovering abiding structures of crime and ethnicity with crucial integrity; one need only think of what Hegel drew from *Antigone* or Freud from *Oedipus Rex*. These works have always worked as informants but they were nobody's fools; they talked to the philosophers because they had an inside knowledge. So literature, which is by no means an innocent bystander but often the accused, a breeding ground of hallucinogenres, has something to teach us about ethical fractures and the relationship to law. Gustave Flaubert's book went to court; it was denounced as a poison.[10] A work, no matter how recondite, specialized, or antiquarian, manifests a historical compulsion.

Of course, we no longer exist in a way that renders manifestation possible: we have lost access to what is manifested and even to manifestation itself. Nothing, today, can be manifested, except, possibly, the fact that humanity is not yet just, the indecency of a humanism that goes on as if nothing had happened. The task of extremist writing is to put through the call for a justice of the future. Henceforth, justice can no longer permit itself to be merely backward looking or bound in servility to sclerosal models and their modifications (their ''future''). A justice of the future would have to show the will to rupture. ''A thinker,'' Flaubert has said, ''should have neither religion nor fatherland nor even any social conviction. Absolute scepticism.''[11] Radically rupturing, the statement is not merely subversive. It does not depend upon the program that it criticizes. How might one free oneself from the ineluctable cowardliness pressing upon social convictions of the present, subjugated as they are to reactive, mimetic, and regressive poses?

On generalizing the notion of addiction, our ''drugs'' uncover an implicit structure that was thought to be one technological extension among others, one legal struggle, or one form of cultural aberration. Classifiable in the plural (*drugs* being a singular plural), they were expected to take place within a restricted economy. What if ''drugs'' named a special mode of addiction, however, or the structure that is philosophically and metaphysically at the basis of our culture?

It has been said, convincingly I think, that the pervert does not do drugs. Perhaps this refers to actions that are executed with guiltless precision. On the noncontingency of addiction, leaving aside the more obvious examples, we also have ''proper'' names: Proust (cortisone abuse); Walter Scott, Charles Dickens, Elizabeth Barrett Browning (frequent recourse to laudanum); Novalis, Kleist, Wackenroder (''soft'' drugs); Balzac (coffee). Do these not point to the existence of a *toxic drive*? The need to ensure a temporality of addiction? The history of our culture as a problem in narcossism?

Crack disappoints the pleasure a drug might be expected to arouse. Hence, the quality of crack as pure instance of ''being-on-drugs'': it is only about producing a need for itself. If Freud was right about the apparent libidinal autonomy of the drug addict, then drugs are libidinally invested. To get off drugs, or alcohol (major narcissistic crisis), the addict has to shift dependency to a person, ideal, or the procedure itself of the cure.

Thomas De Quincey's *Confessions of an English Opium Eater* can be shown to perturb an entire ontology by having drugs participate in a movement of unveiling that is capable, however, of discovering no prior or more fundamental ground.[12] Unveiling and unclouding, opium, in De Quincey's account, brings the higher faculties into a kind of legal order, an absolute legislative harmony. If opium—what elsewhere can be called "tropium" because of its rhetorical rush in his text—perturbs ontology, this is in order to institute something else. The ontological revision that it undertakes would not be subject to the regime of *aletheia*, or rather, the clarity that opium urges is not dependent upon a prior unveiling. Where the warring parts of the *Confessions* refuse to suture, one detects the incredible scars of decision. Always a recovering addict, Kant's subject, in the *Anthropology*, was not particularly pathological in the pursuit of his habits; De Quincey's addict has been exposed to another limit of experience, to the promise of exteriority. Offering a discreet if spectacular way out, an atopical place of exit, drugs forced *decision* upon the subject.

Self-dissolving and regathering, the subject became linked to the possibility of a new autonomy, and opium illuminated in this case (Baudelaire, though under De Quincey's influence, was to use it differently) an individual who finally could not identify with his ownmost autonomy but found himself instead subjected to heroic humiliation in the regions of the sublime. Opium became the transparency upon which one could review the internal conflict of freedom, the cleave of subjectivity where it encounters the abyss of destructive *jouissance*.

The ever-dividing self was transported on something other than the sacred, though the effects of revelation were not unrelated. Decisions would have to be met, one had to become a master strategist in the ceaseless war against pain. The most striking aspect of De Quincey's decision resides in the fact that it resists regulation by a telos of knowledge. To this end his elaboration has critically uncovered for us a critical structure of decision to the extent that it has been tinctured by nonknowledge, based largely upon a state of anarchivization.[13] This leaves any future thinking of drugs, if this should be possible, in the decidedly fragile position of system abandonment. There is no system that can currently hold or take "drugs" for long. Instituted on the basis of moral or political evaluations, the concept of drugs cannot be comprehended under any independent scientific system.

These observations do not mean to imply that a certain type of narcotic supplement has been in the least rejected by metaphysics. To a great degree, it is all more or less a question of dosage (as Nietzsche said of history). Precisely because of the promise of exteriority that they are thought to extend, drugs have been redeemed by the conditions of transcendency and revelation with which they are not uncommonly associated. Qualities such as these are problematic because they tend to maintain drugs on "this side" of a thinking of experience. Sacralized or satanized, when our politics and theories prove still to be under

God's thumb, they install themselves as codependents; ever recycling the transcendental trace of freedom, they have been the undaunted suppliers of a metaphysical craving.

There can be no doubt of it. What is required is a genuine ethics of decision. But this in turn calls for a still higher form of drug.

What follows, then, is essentially a work on *Madame Bovary*, and nothing more. If it were another type of work—in the genre of the philosophical essay, a psychoanalytic interpretation, or a political analysis—it would be expected to make certain kinds of assertions that obey a whole grammar of procedure and certitudes. The prestige and historical recommendation of those methods of inquiry would have secured the project within a tolerably reliable frame. However, it is too soon to say with certainty that one has fully understood how to conduct the study of addiction and, in particular, as it may bear upon drugs. To understand in such a way would be to stop reading—it is to close the book, as it were, or even to throw it at someone. I cannot say that I am prepared to take sides on this exceedingly difficult issue, particularly when the sides have been drawn with such conceptual awkwardness. I would venture the hypothesis that it is as preposterous to be "for" drugs as it is to take up a position "against" drugs. Provisionally they may be apprehended as master object of considerable libidinal investment, whose essence still remains to be determined.

Under the impacted signifier of drugs, America is fighting a war against a number of felt intrusions. They have to do mostly with the drift and contagion of a foreign substance, or of what is revealed as foreign (even if it should be homegrown).

Like any good parasite, drugs travel both inside and out of the boundaries of a narcissistically defended politics. They double for the values with which they are at odds, thus haunting and reproducing the capital market, creating visionary expansions, producing a lexicon of body control and a private property of self—all of which awaits review. Drugs resist conceptual arrest. No one has thought to define them in their essence, which is to say "they" do not exist. On the contrary. Everywhere dispensed, in one form or another, their strength lies in their virtual and fugitive patterns. They do not close forces with an external enemy (the easy way out) but have a secret communications network with the internalized order. Something is beaming out signals, calling drugs home.

To the extent that addiction was at one point within the jurisdiction of *jouissance*—indeed, we are dealing with an epidemic of misfired *jouissance*—the major pusher, the one who gave the orders to shoot up, was surely the superego. In order to urge the point with some sustainment of clarity we shall have to enter the clinic of phantasms that Flaubert chose to call *Madame Bovary*.

After the explosion, there were only a few things left. The refrigerator apparently had been used as a strongbox. That's where I found these papers. They seemed to be very old, dating from the 1990s. I admit that what I saw in that refrigerator sickened me. The papers dealt with what was then called a "war on drugs." They were investigating a female subject, a certain Emma Bovary, who seemed to be a foreigner. Since Bovary was out of jurisdiction or from another era, they'd had a hard time proving she was dealing drugs. In those days they didn't yet know that good drugs were always haunted or contaminated by bad drugs.[14]

Emma Bovary had busted a logic of reappropriation, collapsing the dreams of restoring a self. Drugs, a kind of circulating nonessence that originates in a foreign market or at home, hadn't been invented yet, but their concept was in place, in some place called Rouen. Not that that's where it all started—it had never not started. By the 1990s, however, they declared a war on "these artificial, pathogenic and foreign aggressions."[15] There was one document, supplied by a local authority, saying the technology of drugs responded to the call of addiction. This document seemed unfinished, so I left it out. Someone else wrote that everything he had said about technology can be applied to drugs: acceleration, speed, inertia, the third interval.[16]

It seems that the ideals espoused by medicine and the addict were the same: to deaden the pain and to separate from a poisonous maternal flux. Emma Bovary was apparently a grand self-medicator. Like others before her, she experienced the dangers of a belle âme: raptures that cut her off from reality, hallucinated plenitude and pure communication, a kind of hinge on transcendental telepathy. Everything she tried out—religion, reading, love rushes, getting dressed in the morning—had hallucinogenic, analgesic, stimulating, or euphorizing effects upon her. She would also experience tremendous crashes. The peak of drugs, as of love, was for her, telepathy, a communication over distances. She demanded hallucinatory satisfaction of desire in a zone that no longer distinguished between need and desire. This Emma Bovary ran the paradoxical circuit of self-conservation. The circuit was installed when she made the discovery of God's insufficiency. Without the paternal metaphor holding things together, one was at a loss, one became the artisan of one's own body, fiddling around, experimenting, creating new parts or treating the psyche like an organ, a sick organ. One became a maniacal *bricoleur* of one's own body! It wasn't clear then whether the body was private property or not, whether the authorities could legislate zoning ordinances, or whether pleasure and liberty were values freely exercised upon a coded body. Shit was happening. God's fundamental breakdown, His out-of-serviceness and withdrawal from the scene, meant that she had to replace the emptiness with a symbolic authority. That's when the panic set in, the emergencies that invaded the entire scene. Nothing else mattered; she needed her dose, and she started responding like an addict to the alarm signals that proliferated around her. She didn't care what she took. Hard or soft drugs were an opposition established by medical and legal institutions. The Other was devastated, without address. Who were you going to call upon or appeal to? She would have to mime another plenitude. She started elaborating this Other as absence, and began her work of producing secrets. Alternately proud and anguished, above all secretive, she learned that the formation of secrecy engages a relation to absence and separation; she was working over the etymology of the signifier, trying to recover the substance that was separated from her. She started working overtime, making an orifice operate in its relation to emptiness and to time. It was like a narcissistic overinvestment of an organ, but an organ that wasn't merely just that. Like Glover had said, any substance can function as a drug. One thing was sure: the addict was working what was wanting, missing. Freud had once claimed, in *Civilization and Its Discontents*, that pain is imperious, obeying only the action of the toxin that suppresses it. It seems pretty clear from these papers that Emma Bovary never did any real work. Failing to make any responsible effort, she also failed to meet the requirements of an authentic alterity. She was more into forgetting and the simulacrum than she cared for truth. This would never be forgiven. It didn't matter whether what she did was comestible, smokable, or shootable; she was a hallucinator, a creature par excellence of the simulacrum.

Emma Bovary was executed by a disastrous economy of painkillers. Her fictions quickly turned into devouring creditors. They were submitted to an intensified law of supply and demand, the *suppléance* of an addict's knowing body. It is a knowing related to unknowing, the massive distress signals from beyond any pleasure principle.

But what were these papers doing in a refrigerator in Alameda County? It must have had to do with the fact that America was taking suicide pills in those days, spreading the Jonestown effect on a rebound from exile. America still had to take drugs seriously, to stop *using* them for brute and primitive ends. The right wing's dependency on drugs was well known. No one knew how to disrupt the power of legal prescriptions that continued unthinkingly to make claims about substance abuse and the metaphysical subject. There were some commandos of residual resistance. They were wondering, Where can thinking take place so it would be dispensed to the poor, the body-broken, the racially hallucinated other? They figured, drugs, this nonessence, had to be submitted to thought and not merely legislated out of the political body—and what a body it was! Plugging orifices, building muscle. Still, I don't know what Emma Bovary was doing in that refrigerator. It should have been a more direct hit, like Burroughs, Artaud, Michaux, or maybe something from the American drunkyards, like Poe, Faulkner, London, Chandler, Hammett, or Acker. As I read the documents I realized that she was the body on which these urges started showing almost naturally, prior to the time the technological prosthesis became available on the streets and drugs had become an effect of institution, convention, law. She hit the streets, too, roaming around, breaking out in cold sweats, hiding under veils. The whole addiction thing was a kind of veil, covering over and enshrouding her face, drained and sallow. She was under pressure to a body, but this meant destroying her own body. "*Her own body*,'' what a joke! All sorts of forces inscribed this body like so many invisible graffitti. She declared war on the real, this unknown horror, she put out the call for a drug culture. She worked out of her own abysses, hunting down the imaginary phallic supplement. It took me quite a while to decide on publishing these papers. I haven't shown them to anyone, in part because I don't particularly want to be associated with this discovery. Despite everything, Emma Bovary had had a pure body, she belonged to an unpolluted era of literature. Of course, this was a phantasm. No one really believed it. But still, they wanted to keep it clean. They were straight. The horizon of drugs is the same as that of literature: they share the same line, depending on similar technologies and sometimes suffering analogous crackdowns before the law. They shoot up fictions, disjuncting a whole regime of consciousness. Someone said once that literature, as a modern phenomenon, dating from the sixteenth or seventeenth century, was contemporaneous with European drug addiction. When I saw that refrigerator, I knew there was going to be trouble. I had gone through a lot myself recently—problems with eating or sleeping, and then this house was blown apart. I took the papers and started reading them that night. A sympathy developed for Emma Bovary and even her investigators. I couldn't stop reading, it was like I was becoming these persons. All forms of identification that are structuring emerge from a trauma, or from a reserve of what is missing. I knew that much. So now I want to go public with these papers. I'm not exactly sure why, maybe because back then, someone started thinking about drugs before the place blew up. I hope you don't mind going through this again. It won't be easy. I ask the telereader to become a friend to Madame Bovary, to spend the night with her. It was another, agonizing night of withdrawal.

On Ek-static Temporality

The structure of addiction, and even of drug addiction in particular, is anterior to any empirical availability of crack, ice, or street stuff. This structure and necessity are what Flaubert discovers and exposes. A quiver in the history of madness (to which no prescription of reason can be simply and rigorously opposed), the chemical prosthesis, the mushroom or plant, responds to a fundamental structure, and not the other way around. Of course, one can be hooked following initiation and exposure but even this supposes a prior disposition to admitting the injectable phallus. What do we hold against the drug addict? A mysticism in the absence of God, a mystical transport going nowhere, like the encapsulated carriage of Emma B. It is possibly of some importance that a flower of a different sort, a hallucinated woman, be made to experiment with what we can still call the transcendental street drug—or with feminine incorporations of a phallic flux. A strong concept of purity shot through with virility will come to dominate the history of Madame Bovary, who bears ''a pince nez which she carried, like a man, tucked in between two buttons of her bodice."[17] Any way you look at it, Emma Bovary carries the marks of her many incorporations of a foreign body. We have yet to grasp the male sex she carries with her, for Emma is not a simply gendered woman. Her prime injections of a foreign body follow the multiple lines of an interiorizing violence.

In the first place, Emma's moments of libidinal encounter are frequently described as experiences of intoxication. The second place, however, may be of more interest. In the second place, therefore, we discover that drugs, when submitted to Flaubert's precision of irony, are after all not viewed as a conduit of escape but at the base of life: an evening with her husband, Charles Bovary. "How could she get rid of him? He seemed to her contemptible, weak and insignificant. How could she get rid of him? What an endless evening! She felt numb, as though she had been overcome by opium fumes. Emma judges Charles to be a weak man."[18] The judgment passes from a position of feminine virility. Emma judges Charles contemptible. His nullity, overwhelming, turns her into a hit man. From the sense of the deadening infinitude of this confrontation, the threatening limitlessness of what is mediocre, Emma reconstitutes existence as an effect of an overdrawn downer. Not only does this passage argue for the refinement of difference—this opiate acts differently from other insinuations of her substance/husband abuse—but it shows the opium base to be at the bottom of life. Life in its essential normalcy (they are at the dinner table when she ODs) yields to death because it is on the side of an endlessness that numbs. And so Emma Bovary's body gets rigid with the pressentiment of nothingness. Like the Western world, there is no place or moment in the life of Madame Bovary that could be designated as genuinely clean or drug free because, being exposed to existence, and placing one's body in the grips of a temporality that pains, produces a rapport to being that is addictive, artificial, and beside oneself. The history of mood, or aesthetic theory, from Baumgart to Heidegger deals ecstasy (Nietzsche: *Rausch*), zoning out (Schopenhauer), inspired trance (Kant). But Emma is only a rookie, trafficking in abstract forms of forgetting. She suffers endlessly from her finitude.

Finitude encompasses not only the limited life span of her body but also the physical limitations to which that body, *during life*, is subjected. By now we have traveled along the melancholic contours of this body to its addictive recesses. The body of the addict invents a supplementary organ that discloses itself particularly in moments of abstinence, the negative complement to a fiction of immediate satisfaction. The addicted subject radically encounters the phantasm of lack in abstinence. It has fabricated and textured an organ implant that requires absolute attention in the mode of care. Thus abstinence aggravates and accelerates the relation to lack, disclosing terrific contours. There's no giving up the other. The body of the addict, engendering dependency and the possibility of a *chemical prosthesis*, withdraws from the nostalgia of the body's naturalistic/organic self-sufficiency.

What is Emma, literary philter, on to? On the one hand, drugs are linked to a mode of departing, to desocialization—much like the activity of writing, to the extent that it exists without the assurance of arriving anywhere. Considered nonproductive and somehow irresponsible, a compulsive player of destruction, being-on-drugs resists the production of value. Emma exists somewhere between the drug addict and the writer (she is a writer—at least, she owns the equipment, the stylus). Obsessed and entranced, narcissistic, private, unable to achieve transference, the writer often resembles the addict. This is why every serious war on drugs comes from a community that is at some level of consciousness also hostile to the genuine writer, the figure of drifter/dissident, which it threatens to expel. Like the addict, this writer is incapable of producing real value or stabilizing the truth of a real world. The differences between them are not difficult to discern, and yet a single logic of parasitism binds the two activities to each other. The drug addict offers her body to the production of hallucination, vision, or trance, a production assembled in the violence of nonaddress. This form of internal saturation of self, unhooked from a grander effective circuit, marks the constitutive adestination of the addict's address. Going nowhere fast, as we say, Madame Bovary in this regard signs up for the drug program to the extent that she resumes the violence of nonaddress. ''She had bought herself a blotter, writing-case, pen-holder, and envelope although *she had no* one to write to'' [Elle s'etait acheté un buvard, une papeterie, un porte-plume et des enveloppes, quoiqu'elle *n'eût personne à qui écrire*].[19]

With nowhere to go and little to do, these missives, along with the equipment that maintains them, can only be routed inwardly. But it is an inwardness of diminished interiority, a kind of dead letter box—an impasse in destination. Still, writing for no one to no address counts for something; it is the writer's common activity. For Flaubert, this movement of the simulacrum without address (or, in another idiom, without purpose, point) is associated with the toxic pleasure of a certain narcissism: I have condemned myself to write for myself alone, for my own personal amusement, the way one smokes or rides.[20]

It is important to weigh this violence of nonaddress because it designates a most vulnerable type of writing that is, like smoking, susceptible to acts of nihilism, burning out. Unaddressed or unchanneled pleasure, this condemnation to solitary confinement, with or without a community of smokers, belongs to the registers of a ''feminine'' writing in the sense that it is neither phallicly aimed nor referentially anchored, scattered like cinders. At no point a prescriptive language or pharmacological ordinance, it is rather a writing on the loose, running around without a proper route, even dispensing with the formalities of signing. The impropriety of such writing—which returns only to haunt itself, refusing to bond with community or affirm its health and value—consistently reflects a situation of depropriation, a loss of the proper. Thus the heroine (who is also, sometimes, Flaubert: "Madame Bovary, c'est moi!") not only has one to write to, but also lacks a proper name: " 'Madame Bovary! Everyone calls you that! And it's not even your name—it's someone else's . . . someone else's!' He buried his face in his hands."[21] Still, this is the name that entitles the book, and cosigns its cover.

Going up in Smoke

This reading does not seem to accord with literary criticism in the traditional sense, yet it is devoted to the understanding of a literary work. Perhaps it resides within the precincts of philosophical endeavor, for it tries to understand an object that splits existence into incommensurate articulations. This object resists the revelation of its truth to the point of retaining a status of insurmountable otherness. Still, it has given rise to laws and to moral pronouncements. This in itself is not an alarming occurrence. The problem is signaled elsewhere, in the exhaustion of language. Where might one go today, to what source can one turn, in order to activate a just constativity? We no longer see in philosophy the ultimate possibilities for knowing the limits of human experience.

And yet I began this essay by citing Friedrich Nietzsche. There were two reasons for this selection. In the first place, Nietzsche was the first philosopher to think with his body, to "dance," which is a nice way of saying also to convulse, and even to retch. And then, Nietzsche was the one to put out the call for a supramoral imperative. This in itself will urge us on—for we are dealing in a way with the youngest vice, still very immature, still often misjudged and taken for something else, still hardly aware of itself.

Notes

1. Martin Heidegger, *Being and Time*, trans. John Macquarrie and Edward Robinson (New York: Harper and Row, 1962), sec. 41.

2. Jean-Luc Nancy, *The Inoperative Community*, ed. Peter Connor (Minneapolis: University of Minnesota Press, 1991); Maurice Blanchot, *The Unavowable Community* (Barrytown, N.Y.: Station Hill, 1989).

3. Ernst Jünger, *Annäherungen: Drogen und Rausch* (Stuttgart: Ernst Klett Verlag, 1970).

4. Many of the assertions made here are elaborated in Avital Ronell, *Crack Wars: Literature/Addiction/Mania* (Lincoln: University of Nebraska Press, 1991).

5. Friedrich Nietzsche, *The Gay Science* (New York: Penguin, 1986), 132.

6. Charles Baudelaire, *Les paradis artificiels* (Paris: Gallimard, 1961). See also Théophile Gautier, *La Pipe d'opium, Le Hachich, Le Club des Hachichins* (Paris: Gallimard, 1961).

7. Heidegger, *Being and Time*, 195.

8. Ibid., 195.

9. Martin Heidegger, "Das Seins des Daseins als Sorge," in *Sein und Zeit* (Tübingen: Max Niemeyer Verlag, 1986), 191.

10. See also Dominick LaCapra, "*Madame Bovary*" *on Trial* (Ithaca, N.Y.: Cornell University

Press, 1982), 38. LaCapra, discussing the position of prosecuting attorney, Ernest Pinard, concludes, "The novel is literally poison."

11. Letter to Victor Hugo, cited in Francis Steegmuller, *Flaubert and Madame Bovary: A Double Portrait* (Chicago: University of Chicago Press, 1977), 276.

12. Thomas De Quincey, *Confessions of an English Opium Eater* (1822; London: Penguin, 1971), 71.

13. I refer to Jacques Derrida's coinage of "anarchivization" in his contribution to the colloquium "Lacan avec les philosophes," organized by René Major and Philippe Lacoue-Labarthe, Paris, May 1990.

14. See Jacques Derrida's discussion in "Plato's Pharmacy" in *Dissemination*, trans. Barbara Johnson (Chicago: University of Chicago Press, 1981).

15. Jacques Derrida, "Rhétorique de la drogue," in *L'Esprit des drogues: La Dépendence hors la loi?* ed. Jean-Michel Herviev (Paris: Autrement Revue, 1989), 30.

16. Paul Virilio, seminar offered at the Collège International de Philosophie, "Téletopies," March 5, 1990.

17. Gustave Flaubert, *Madame Bovary*, ed. Leo Bersani, trans. Lowell Blair (Toronto: Bantam, 1981), 4.

18. Ibid., 217.

19. Ibid., 52; emphasis added.

20. Cited in Steegmuller, *Flaubert and Madame Bovary*, 247.

21. Flaubert, *Madame Bovary*.

PART II
Technology and the Environment

Chapter 5

Eco-Subjects

Verena Andermatt Conley

No more Shakespeare after Chernobyl.
Jean-Luc Godard, in *King Lear*

The only struggle worth fighting for is a truly ecological struggle.
Paul Virilio, *Défense populaire et luttes écologiques*

Eco-Subjects

An *eco-subject* can be defined as the citizen who ''pitches in'' to save the environment. He or she gains self-respect and commands pride when making pilgrimages on foot, bicycle, or (less preferably) by car to neighborhood recycling bins. To gain a temporary sense of communality, the eco-subject separates and piles up cans, bottles, plastic containers, or newspapers and tosses the neatly arranged bags into duly marked dumpsters. Such gestures make us all feel like truly virtuous, good, clean citizens. Yet, as the international conference at Rio de Janeiro on the state of the world fast approaches, we know that this is not just ''what it is all about.'' Ecology has been studied primarily in areas of biology, meteorology, geography, and demography. Less has been said on the subject in the humanities, where its mention is generally parenthetical. It comes as the last and least term in a socioeconomic series of race, class, gender, and ecology. Until recently, problems of the subject have been studied mostly in abstraction. In the wake of romantic subjectivity that follows a Newtonian or mechanistic concept of nature, many theories still advocate putting humankind on the center stage of the world. Michel Serres puts it thus:

> Remove the world around the struggles, keep only conflicts and debates, dense with men, purified of things, you will have the theatrical stage, most narratives and philosophies, all of the social sciences: the interesting spectacle we refer to as ''cultural.'' Whoever says where the master and the slave are struggling?

Our culture cannot stand the world.[1]

Despite its overgeneralizing, the declaration holds. It was confirmed by the recent war in the Persian Gulf, where news of massive (and voluntary) environmental damage was quickly shunted aside in favor of victory and human glory. Serres notes also that, today, "global history enters into nature, global nature enters into history. This is an *inédit* in philosophy."[2] In other words, because of technological advances, we have been able to see the planet in its entirety, and have become aware of global interdependencies of many kinds. If, in addition, we admit that human societies are in constant change, and that every state of "being" is but the effect of a temporary historical configuration, we can no longer think the subject, singular or collective, in a vacuum. At the very moment when technology has seemingly acquired mastery of nature, "nature" returns and is rehistoricized. Henceforth, human subjects, always in movement and transformation, have to think themselves in a world in becoming. This new nature that now includes the biosphere points to global interconnections of forces, influences, causes, and effects that have no national boundaries. In the pages to follow, I wish to sketch briefly the rudiments of a program that deals with the effects of such a double becoming and/on its subject.

Eco-Feminism

Earlier, before *ecology* acquired its current usage, feminists such as Hélène Cixous, Luce Irigaray, and others insisted on the necessity of thinking a human subject *in* the world. They sought to decenter man's self-designated role as master of the world, a position that, they argued, had led to oppression of women, colonialism, and genocide. They felt that "man" is linked to what can be called the "Occidental project"[3] that reaches far beyond the historical and geographical occupation of the Third World by a First World after the age of discovery. At the confines of psychoanalysis and philosophy, the same feminists sought ways to transform Hegelian dialectics and to keep the other—any other—alive. They did so in gesture and in writing, through programmatic changes they both advocated and performed in language and representation. They argued for a decentering and a setting in motion of the self and complicated the concept of being through becoming. Without explicitly calling it such, they argued for singular and collective ecological relations to the world. For the purpose of the argument that follows, I will see how these feminists' pronouncements can serve to develop an ecological component and where and how they need to be complicated.

In the wake of Heidegger's writings distinguishing between two technics, an instrumental technics of enframing and a *techné* as *poiesis*, Hélène Cixous writes with lyrical effusion:

> I have an Oranian childhood that remembers the plants at the bottom of the hill in the *Jardin d'Essais*. What I can still understand of the language of plants, I learned from childhood. It was a childhood completely faithful to the world. . . . It is the childhood that still knows in me, how to be in a garden without distancing it . . . : without trying to appropriate it. . . .
>
> We have been taught the language from above, from afar, that listens to itself . . . that understands only in translation; speaks only in its language, listens only to its grammar; and it is because of its order that we are separated from things.[4]

The adult subject, she implies, no longer knows how to hear the language of things. Their idiom has been covered with a magma of intellect that ideologizes and fixes the subject, who tries not to listen to them but to categorize their constitutive parts. Access to symbolic laws comes through apprehension of grammatical rules. How, Cixous asks, can we still hear the language of things across these barriers? Feminism offers a means to do so. It does not advocate power reversals, but devises ways of letting both others (humans) and other "things" (organic and inorganic) merely be. The division into subject and object, the post-Cartesian operation that reduces and flattens the world through use of a language detached from the body, is said to lead to repression. By contrast, in accord with the "good" Mother Nature and her mythological representative, Demeter, Cixous carries in herself the world as a vast childhood memory. This enables her to be in tune or accord with the world, to hear the language of things.

When we are forced to obey uncritically a repressive system of signs that makes up a symbolic order, we lose our affective contact with the world. Cixous makes it clear: The Occident's way of using language—what Derrida elsewhere calls phallogocentrism—advocates takeover of people *and* things. She notes the generalized loss of the world resulting from a constrictive practice of language from which she excludes herself. At the same time, she advocates the necessary move of letting the world of things be or of approaching them with tact, the first step toward an ecological rapport. If technology is destined as modes of the management of natural forces, it can be asked if the desire to let be is not a longing for the impossibility of a pretechnological world. We can also wonder—with Jean-Luc Godard—if the lyrical encomium of the writer as Mother Nature is adequate at a time when the world is threatened by massive ecological catastrophe. Like other feminists, Cixous writes of a subject in constant change and motion, in brief, in becoming. She refuses to be caught in the rigid mold of a *techné* that serves a dominant patriarchal model. By opting—in a Heideggerian mold—for another *techné* as *poiesis*, she indeed urges an ecological link between subject and world. Yet her stance is resolutely antitechnological. By refusing to deal with technology in terms other than the process of "writing" or "representation," she willfully decides not to look at many of the interconnections of the contemporary world. Though she herself avowedly makes a shift in her thinking from the subject of the unconscious to that of history, she does not complicate her concept of nature. For feminists, like Cixous, the world of things, or nature, is lost through symbolic language and grammar. To let things speak, we have to have recourse to presymbolic language. For her, acceleration through technology, the media, and symbolic language deadens our senses. We become by separating from the grid and yoke they impose on us.

Our critical task entails seeing how we can complicate Cixous's notion of a subject in the world by juxtaposing it to that of other thinkers who do incorporate instrumental technology and a redefinition of nature in order to define programs for ecological criticism. I have chosen to look at passages from several cultural theoreticians whose work touches upon relations between ecology and technology, either positively or negatively. I will evaluate them briefly and test them against feminist findings to see how we can sketch an outline for a general program of a cultural ecology and elements for an ecological subject.

Echoing Paul Celan's commentary about the future of poetry after the momentum of genocide, in a mixture of parody and gravity, Jean-Luc Godard declares in his filmed version of *King Lear*: "No more Shakespeare after Chernobyl!" Paul Virilio asserts in *Défense populaire et luttes écologiques*: "The only struggle worth fighting for is a truly ecological struggle."[5] Each remark begs for a return to Heidegger's searching and questioning: What form of technics will help displace or resist the devastating effects of instrumental technology that has led to irretrievable loss of species and habitat? How can one attempt to deal with the overwhelming magnitude of human suffering in the famines that accompany demographic explosion? What theories will inform those who have to relinquish some of their goods and the have-nots, who also must eventually carry their share of responsibility?[6] Are there, if one of Cixous's theoretical terms can be used, unknown *sorties* (ways out) of the dilemma, or are we the powerless witnesses who can only utter cries of rage at every bit of news: demographic explosion, threats to species, the disappearing ozone layer, worldwide famines. Although I do not wish to refute other, more "archaic," relations to the world that make the subject of poetry, I do not simply want to ignore the world around me. Is it possible to devise rapports that, for being ecological, will not entirely reject contemporary technology?

Ecology/Technology

Ecology is not just another transcendental signifier, but a concern traversing *all* political and emancipatory discourses.[7] Subjects have to be thought—whether in a singular or collective sense—*in* a nature redefined in terms of fluctuations, patterns, and randomness rather than of Newtonian immobility and timelessness. Cixous launched her critique of the phallocentric, full subject with a feminine subject in movement and transformation. A feminine mode of exchanging let the self be altered through its tactful *écoute* of the world. Technosciences have added to this poetic model others that show that nature too is in transformation or becoming, and that, under stress, unwarranted alterations are prone to occur. Rapid transformations under the impact of technology make societies volatile and ready to bifurcate. Nature, too, under pressure, may at any time undergo irreversible changes. Becoming henceforth revolves around both human beings and nature. Extension of the definition requires us to rethink concepts of dynamic equilibrium and human activity in terms of global interrelations. Under the pressures of variously webbed relations, utopian views of harmony or romantic notions of the limitless are being rewritten. New limits—or thresholds—alter our ways of dealing with social and natural ecology in at once local and global terms.

Acceleration does not only institute a linguistic separation between humans and things. David Harvey shows how a dominant system actually brings about loss of place and space, in both physical and metaphorical senses. As a geographer, Harvey is more attuned to actual space and asks how we can recoup it. He too urges for an opening through becoming.

The geographic history of capitalism, related to the shrinking of the globe through innovations in transportation and telecommunications, led to what have been called time/space compressions. As Harvey puts it: "A strong case can be made that the history of capitalism has been characterized by speed-up in the pace of life, while so overcoming spatial barriers that the world sometimes seems to collapse inwards upon us."[8] Acceleration accentuates globality of biological interdependencies. Space appears to shrink, as Harvey says, to "a 'global village' of telecommunications and a 'spaceship earth' of economic and ecological interdependencies—to use just two familiar terms and everyday images—and . . . time horizons shorten to the point where the present is all there is and we have to learn to cope with an overwhelming sense of compression of our temporal and spatial worlds."[9] Yet acceleration is also linked to increased economic exploitation. Ecological and economic problems are inextricably intertwined. Integrated world capitalism (IWC), or market economy, based on acceleration enabled by means of high technologies, has led to unprecedented wealth in some areas and hardship in others.[10] Vandana Shiva, a former nuclear physicist from India turned ecologist, confirms that

> Northern Speculators . . . "gamble" not only with the wealth of nations but also with the lives of powerless farmers within those nations. Wealth from the South is transferred to the North in a new wave which colonizes the land and forests of the Third World through commodity prices and futures markets. Entire countries, ecosystems and communities are vulnerable to instant collapse in this game of speculation, which bids on them and their produce, and then abandons them as waste—wastelands and wasted people.[11]

The acceleration of IWC has wrecked the economies of many ecological—that is, from a Western point of view, ''backward''—societies. The acceleration and implosion of our world through high technology is intricately linked to an ideology. The motto ''The present is all there is'' disavows the past and refuses to consider the future of things. Under the guise of a *fake becoming*, that is, of a betterment and irreversible improvement of reality used for the marketing of products, capitalism thrusts people forward while engaging them in a deadly race in which profit is the only motive for advancement. It wipes out, in the name of social engineering and so-called new world orders, forms of human or animal life that go back thousands and thousands of years.[12]

Ecological societies, or other societies based on more constant time, low technologies, and various archaisms such as mythologies, astrologies, and mysticism, are thought to be hopelessly ''backward''—except in feminist fictions—in relation to an international capitalist system. Harvey tries to be critical by asserting the necessity of a real ''becoming,'' that is, of tracing a way out of the postmodern time/space compression understood as a historical-geographical condition. How can we, in culture, recoup a lost sense of place or space? He sees the possibility of a counterattack of narrative against image and information. Although Harvey at times appears swept up in the whirlwind of the present, in the vertiginous rush of accelerating technologies sold by advertisements that we endow with an a priori notion of progress, he brilliantly denounces what he calls ''flexible capitalism'' by pointing toward the deterioration of living and social conditions under the influence of shrinking job markets and skyrocketing demographics (perhaps without sufficiently underscoring the exploitative nature of a technological society in the era of IWC and resulting problems in natural and social ecology). Shiva appears to cast her words in a similar way:

> The survival of the poor and the future are being sacrificed to keep the casino [society] running. . . . In terms of scale and sheer magnitude, the tribute extracted from the Indian subcontinent (and one of the major sources of financing the eighteenth century industrial revolution) by such nabobs as Warren Hastings and the British East India Company, pales in comparison to the current outflows.[13]

Ecology calls for a transformation in both theory and application of scientific technology.

To repeat, after extensively dwelling on the development of glitzy media art in the 1980s, such as architecture and painting, Harvey makes a somewhat quick detour through film, and especially narrative, in order to argue for resistance. For Cixous, poetry—and storytelling—allowed the subject a way out of the rigid symbolic grid that had been imposed on her and opened onto becoming, passage, and the limitless. Such a position is complicated here with attention to actual loss of space and place in current conditions of compression that infringe on our conscience.

Celebration of Waste

Limitless perspectives advertised in theories of the avant-garde are called into question by ecological—both natural and social—considerations. Limits go through a world that does not rule out chance. The celebration of boundless energy consumed as *dépense*, as a kind of inverted becoming, no longer operates. To emphasize the point, we need only recall what happens when Georges Bataille's theory of the *part maudite*, or accursed share, is pushed to an extreme. Jean Baudrillard, pursuing an argument reminiscent of Marinetti, resembles the engineer, quoted by Paul Virilio, who delighted in the beautiful spectacle of the *marée noire* in Brittany caused by the running aground of the *Amoco Cadiz*. His aestheticizing pronouncements are daubed with fascistic overtones. Baudrillard asserts that order may come out of disorder, that pattern can emerge through chaotic randomness of ecological disaster; running contrary to classical theories of harmony and equilibrium, he states that new orders may come from extreme fluctuations or conditions of disequilibrium. Delighting in catastrophe that appears to be a precondition to a second coming (so imbued is his belief with biblical prophecy), Baudrillard writes:

> It could be that the entire system of transformation of the world through energy has entered into a viral and epidemic phase, corresponding in fact, to what energy is in its very essence: spending (*dépense*), a fall (*chute*), a differential, a disequilibrium, a miniature catastrophe that first produces positive effects but that, overtaken by its own movement, takes on dimensions of a global catastrophe.[14]

Presumably, a healthy ecology will then *follow* a necessary disaster that man has to impose on the environment.

Enhancing his remarks, Baudrillard takes a detour through Manhattan. Residents of the island, he observes, derive their energy from soot, acceleration, and exhaust fumes—that is, from the delight that an unthinkable human environment offers to its population:

> To dissuade people from such prodigality, waste, *inhuman* rhythm would be a double mistake since they draw (*puiser*) from what would exhaust (*épuiser*) a normal being the resources of an abnormal energy and that . . . they would be humiliated to have to slow down and economize their energy. It would be a degradation of their collective standing, that of an excess (*démesure*) and of an urban mobility that is unique in the world of which they are the conscious and unconscious actors.[15]

Distinguishing risk from *lack* and exhaustion from *excess*, he adds that nothing will be able to stop this internal logic of acceleration of a movement that is already "out of balance."[16] In arguing for a corrective catastrophe, he belittles those who—at the time of chaos theory—still advocate a mechanics of homeostatic equilibrium. Their interests are contained in a "safeguard of the species through ecological conviviality."

Celebrating "the inhuman," a "viral" or "epidemic" phase—which, in view of the worldwide problem of AIDS, which is equally related to ecological problems, takes on a more than cynical ring—Baudrillard ridicules "conviviality" as whatever has naive implications of sociality. He becomes the protoype of the free-based, prophecy-believing capitalist bringing together former theories of the left, a *dépense effrénée*, and the position of the New Right that bears total inattention to global well-being. And he predicts that at the end of this race, another destiny of the human species, and other symbolic relations with the world—more complex and ambiguous than those pertaining earlier models of equilibrium and interaction—will come about: "A vital destination would imply a total risk."[17] Yet market economy takes no risks, at least not intellectual ones. It is interested only in selling, just as Baudrillard is in marketing a model of doom.

A New Alliance

Chaos theory, at the basis of much ecological thinking, does not have to follow Baudrillard's chronology. Certain research in contemporary physics has moved away from theories of immobility, equilibrium, and objective truth. It has reintroduced point of view and uncertainty into the advanced technology and, consequently, has shown that any new implementation of elements in a global system will have to be made with utmost prudence. Chaos theory criticizes linear notions of progress, especially the doublet of *new = better*. It cautions about the unexpected. Some of the very scientists from whom Baudrillard derives his wisdom (unacknowledged), such as Ilya Prigogine and Isabelle Stengers, have shown how the discovery of complex systems reveals the very uncertainty of our thought on the universe. Around the notion of becoming, now operative in both science and humanities, for ethical and ecological purposes, they call for a new alliance between the two. They do so at the very time they suggest that nature has a history and is not simply immobile and to be acted upon. In the conclusion to *Order out of Chaos*, they write:

> It is remarkable that we are at a moment both of profound change in the scientific concept of nature and of the structure of human society as a result of the demographic explosion. As a result, there is a need for new relations between man and nature and between man and man. We can no

longer accept the old a priori distinction between scientific and ethical values. This was possible at a time when the external world and our internal world appeared to conflict, to be nearly orthogonal. Today we know that time is a construction and therefore carries an ethical responsibility.

The ideas . . . of instability, of fluctuation—diffuse into the social sciences. We know now that societies are immensely complex systems involving a potentially enormous number of bifurcations exemplified by the variety of cultures that have evolved in the relatively short span of human history. We know that such systems are highly sensitive to fluctuations. This leads both to hope and a threat: hope, since even small fluctuations may grow and change the overall structure. As a result, individual activity is not doomed to insignificance. On the other hand, this is also a threat, since in our universe the security of stable, permanent rules seems gone forever.[18]

Now, next to this ethical injunction, which does not rule out contingency and uncertainty, Baudrillard's aestheticizing, orgiastic view is applied to a city whose living subjects completely disappear. This totalizing, aestheticizing approach of his signature does not take into consideration the fact that this system of acceleration that he somewhat tendentiously chooses to call *dépense* is one that annihilates countless species that have taken thousands of years to form but that, also, in economic terms (or social ecology), further divides humanity between those who have and those who have not. Notions of instability and uncertainty do not pertain merely to scientific models. They have also been appropriated by capitalism, which advocates that its subjects be "flexible," in other words, change their income and living conditions contingently, but only in order to assure that someone else reaps a majority of profits.

The conclusion of *Order out of Chaos* shows how becoming has superseded being in science as well as in the humanities. Now, to repeat, *becoming*, thought of in emancipatory discourses as an opening, has also been appropriated by IWC for destructive ends. Next to exhorting its subjects, in a minuscule semantic shift, to remain flexible, it mirrors another, false becoming, one of temporal acceleration made possible through technologies, often arrived at by creating obsolescence, hence new markets. Becoming once was linked to undermining the values of bourgeois stability and its symbolic constructs. It was supposed to be liberating. In a market economy, it has shifted toward destruction and the staggering production of unusable trash. This type of fake becoming actually masks a strategic, colonial-type takeover, engineered through the perfection of telecommunications in the sector of those who "have." The operation disperses and immobilizes those who "have not," thus exacerbating local ecological struggles anywhere and at once, all over the globe.

Casting aside Baudrillard, who, through his supercilious tone, perhaps somewhat belatedly wishes to "épater le bourgeois"—that is, to scandalize the middle class—we can repeat, following Prigogine and Stengers, that societies are now viewed as complex systems involving enormous numbers of bifurcations. Highly sensitive to fluctuations, current societal conditions can be viewed as threatening but also as signs of hope. In the context of global interdependencies, local and individual actions may have positive effects. Change is possible—although results may not always be foreseen. But societies are changing in a physical world that has been rehistoricized. In other words, at any social level—whether in Manhattan, the suburbs of Paris, or the rain forests of Surinam—"nature" has reappeared and is also in becoming.

Of Terminal Humans or Human Terminals?

Paul Virilio, at first, would appear to subscribe to Baudrillard's view of the future. Attuned to social contradictions, to collectivities and suffering, Virilio mimes the very system he criticizes by projecting his way of apprehending the present into futuristic visions. Like David Harvey, Virilio is the prophet of capitalism's politics of speed. Fascinated by technology, whose very ideology he condemns, Virilio specifically locates areas of collusion between militarism and capitalism. Whereas Harvey inventories the results of flexible capitalism through statistics and suggests possible links with postmodern forms of expression, Virilio is more specifically concerned with the effects of ecological consequences for the human collectivity. In his writings, abstract developments are increasingly abandoned in favor of denunciation of loss of habitat—in every sense of the term—at the hands of a small industrial elite. He writes at the intersection of a denunciation of capitalism and espousal of ecological struggle. Ecology is conceived both as environment and as a means of coping with the loss of space. The two are closely interrelated. In *Défense populaire et luttes écologiques*, he points to the proximity between the industrial complex and the military. Earlier, he spoke of a *logistique de la perception*, in which he showed the degree to which photographic development is linked to logistics and to military strategy. He called in question a *techno-logistique* that developed technology first and foremost for military purposes. It is the idea of total war that changes human relations to space and time by pushing people out of their habitats while at the same time devastating their environments through chemical means. An earlier *exo*colonization of capitalism, he argues, has given way to an *endo*colonization in which people are being displaced by force. A former *lieu d'élection* (a site of choice) is transformed into a *lieu d'éjection* (a site of ejection) whenever a collective imposition of universal atopia prevents people from *either* being or becoming.

Without providing alternatives, Virilio envisages acceleration as benefiting primarily an invisible, universal, capitalistic power that, through means of telecommunication, produces money—not goods—and is no longer a centralized but a dispersed force. The same logic of acceleration has disrupted social links for greater efficacy and confined people into *habitacles* (tiny living spaces) where they are glued to their chairs, linked by various threads and remote controls to answering machines, VCRs, and computers, and tuned into virtual reality. People are immobilized. Force-fed consumers become part of a renewed spectacle society in which durable products are replaced by ephemeral spectacles. They alternate between gagging on an overdose of images and suffering from withdrawal symptoms. Victim of a market economy, the new citizen is in any case both the terminus and the terminal of a ''teletopic city in accelerated formation.''[19] Virilio insists on the isolation imposed on this mass of consumers who sit in front of the TV set in their *habitacles*, alone or as single parents with their children. We have come far since the earlier feminist critique of the ''bourgeois nuclear family.''[20]

Is there a possible *sortie* from systems exerting such time/space compressions? Despite an almost futuristic passion for technology, Virilio does not see any immediate issue beyond resistance through denunciation. For the purpose of mapping a *practice* of ecological subjectivity, we can juxtapose his hypotheses to those of one of his intellectual peers, Félix Guattari, who also denounces the ecological damage inflicted by IWC. The train of thought of the one can be grafted upon the other. For Guattari, ecologial problems are both metaphorical and ''real.'' Weeds are akin to bad ideas. They pollute the mind and the world.[21] They reduce diversity and can be said to discourage ''rhizomatic'' process on viable surfaces. Guattari's call to action is double. In an existential vein, he calls for a resingularization of our world through aesthetics and, at the same time, for collective action in the context of a generalized *ecosophy*. Singular subjects have to reclaim their right to become but also to be—that is, to live in a habitable world.

Neither condemning technology nor using analyses that currently support the ideology of IWC, Guattari proposes couplings of human and machine that do not oppose organic to inorganic matter. And without rejecting *le fait freudien* (or trashing psychoanalysis on the grounds of its low degree of cost-effectiveness), he sees a liberating potential in the relation between human and computer because, in his way of thinking, it does away with the rigid trinity of earlier psychoanalytic analyses and opens to bifurcations and new becomings. More important for our purposes, the relation with the machine not only dehumanizes people, it alters the very construction of subjectivity. Limits have been pushed back to cosmic bounds where the contrary of reality is no longer simply a dream. Computers plug into virtual universes. At the frontiers of the cosmos, they reveal the very limits and uncertainties of our thought and, paradoxically, bring back the idea that the universe cannot be entirely controlled or its opacity clarified.

Guattari opens the possibility of a future on condition of a total devotion to what he calls "ecosophy," a term laden with mental and social repercussions. Ecosophy would get us out of dialectical oppositions pertaining to antiquated social models. In an era that combines and even equates social and natural struggles on a global scale, dialectics are superseded by notions of bifurcation or rhizomatic process. Among other things, Guattari, like Harvey, calls also via Benjamin for the necessity of storytelling—that is, a way of becoming—as a complement to dissemination of information, or its reduction and repetition. Through the telling of a story, through enunciation, the speaker establishes a *place* and commands a *relation* to time and space. The body is brought in through spacings of speech and through voice. In its broadest context, storytelling does not rule out relations between human and machine, nor does it fall into metaphysical traps of logocentrism. The combination of ecological struggle and the coupling of human and machine would not oppose but displace older social dilemmas. Where Virilio's *homme-terminal* has a sense of an ending, Guattari's promises that of a beginning. Where Virilio focuses on the raw exploitation of the world via advanced technology at the expense of ecosystems, Guattari looks to scientific discoveries that question the very thoughts leading to current imbalances. The question is how to combine the two *hommes-terminal*, that is, how to cope with the threat while combining scientific and humanistic aspects of becoming to provide an opening to the future.

Tact and Diversity

Changes happen quickly, and their contingent solutions make the future difficult to predict. Just as the West claims that the communist utopia has failed, the market economy may also come to an end. Lately, in a tradition of prophecy-belief, much attention has been given to the possibility of another bifurcation as the result of an asteroid hitting the earth and killing most of its life. Yet, before we lose ourselves in such speculations that disregard present actions, we will need to think more of the two *hommes-terminal*. In face of the suffering, past and present, unprecedented because of skyrocketing demographics—still a relatively well kept secret—it is less with a feeling of euphoria than perhaps with shame over a collective genocide and geocide that we may reflect on the past and look toward the future.[22] By way of feminists such as Hélène Cixous, through geographers, philosophers, and culture critics, time and again we hear the need for becoming in cultural contexts of resingularization through storytelling, narrative, or poetry. Through voice, storytelling brings the body, or one's own story, into History. And insofar as it reopens onto space in time, away from technological reduction onto grids, it does preserve linguistic diversity. It also questions the pseudo-objectivity of any truth.

For Jean Baudrillard, a story of apocalypse is the only "master-narrative" available for contemporary life, whereas, as we have observed, infinite possibilities of micronarratives or myth exist when, for Prigogine and Stengers, fluctuations of systems—both local and global—are in dialogue with artistic process. David Harvey suggests the same possibility when he equates local narrative with ecological struggle. Félix Guattari uses "machinic" thinking as a means of making connections among organic and inorganic systems where, formerly, none might have been said to exist (the story of the ionization of a grain of salt in a flowing river could be the beginning of a new and epic *roman fleuve*). For three of the four writers, "modes of singularization" project local practices toward a global future. Their stories and their fantasms share a common and highly material conception of becoming. It is involved with systems that afford a meaning but also, like an unfinished story, a promising degree of unpredictability and randomness. The story is specific insofar as the context in which it works disallows generalization that would resemble a religious vision or master narrative. The practical or means-ends appearance of becoming is only ostensible. Because it uses the history of nature as a point of departure for resingularization, becoming does not equate the present with a zero degree of narrative, but as a complexity that has to be made even more complex, and to be enjoyed not only for the beauty of increasing diversity, but also because it is not subject to totalizing economic rules. And paradoxically, becoming *becomes* when it moves slowly.

At its point of departure, this discussion took up the protoecological condition of feminism, where it was said that poetry is not just verse or song, but any way of thinking an "approach." If we recall Cixous in the context of a generalized ecology, now we can suggest that the disappearance of the world's diversity, its capacity to become, and its sensuous opacity—of legends and narratives—go hand in hand with the wanton cutting of bushes and trees.[23] Through their presymbolic languages, closer to the body, feminists had hoped to develop less hostile relations with the world than those we are witnessing in the regime of market economy. Cixous underlines how the desire to take over is a masculine position.[24] This position has led to the devastation of the planet. By opting for a *double* becoming—that is, for a singular and collective subject in becoming in a world equally in becoming—we see that technology has revealed a position neutral in respect to ecology.[25] Without simply ignoring or simplifying technologies, we now need to emphasize the necessity of thinking the subject not only in its relation with other subjects, but also *in*, and with the astonishing complexities *of*, the world. Yet, paradoxically, at the confines of the universe, contrary to any overriding belief in "man" or "woman," machines show humans that our universe is opaque, or "haunted." Singular narratives do not zap us. And—yet another well-kept secret—becoming that does not eradicate fails to eradicate diversity only when it decelerates.[26] The world, to go back to feminist lessons from which we began, needs to be treated with *tact*, and treated with patience. It is inhabited with linguistic diversity and can *become* when it retains some of the opacities that, no less, have been revealed to us by technology at cosmic frontiers. Yet beyond some limits of the feminist enterprise, there is need (and I shall take up the point elsewhere) to emphasize the necessity to decompress, renarrativize, resingularize to separate oneself from hegemonic points of view, and use multiple tongues that lend the world its diversity, all the while one must be conscious of living in a nature that is, thankfully, being reenchanted. Thus, perhaps, one can hope to open ways that will turn technology toward humans rather than against them.

Notes

1. Michel Serres, *Le contrat naturel* (Paris: Bourin, 1990), 20.
2. Ibid., 19.
3. Edouard Glissant, *Le discours antillais* (Paris: Seuil, 1980), 6.
4. Hélène Cixous, *Illa* (Paris: Des Femmes, 1980), 135-36.
5. Paul Virilio, *Défense populaire et luttes écologiques* (Paris: Galilée, 1978); my translation.
6. Paul Ehrlich, in *The Population Explosion* (New York: Simon and Schuster, 1990), shows convincingly that there are not enough resources to bring the poor up to the level of the rich but that, to the contrary, the rich will have to give up some of their privileges.
7. Glissant, in *Le discours antillais*, seems much aware of the additional problems of emancipatory discourses on a polluted planet.
8. David Harvey, *The Condition of Postmodernity* (Oxford: Blackwell, 1989), 240.
9. Ibid.
10. The expression "integrated world capitalism" was coined by Félix Guattari in *Les trois écologies* (Paris: Galilée, 1990).
11. Vandana Shiva, *Staying Alive* (London: AEC, 1989), 221.
12. For example, in the U.S. Northwest, the building of dams by the Army Corps of Engineers has all but destroyed the migration and spawning routes of the salmon.
13. Shiva, *Staying Alive*, 222.
14. Jean Baudrillard, *La transparence du mal* (Paris: Galilée, 1990), 108. The epidemics seems to be more literal than what Baudrillard implies. Here we can think of the resurgence of cholera and typhoid and, of course, AIDS, which are not solely metaphors.
15. Ibid.
16. It is to be hoped that Baudrillard's social observations are keener than his notions of geography. In *America*, he looks for the Rockies while looking out his Minneapolis hotel window. We may wonder what a French person would say if an American pretended to see the Pyrenees from his Paris hotel!
17. Baudrillard, *La transparence du mal*, 208-9.
18. Ilya Prigogine and Isabelle Stengers, *Order out of Chaos* (foreword by Alvin Toffler) (Boulder, Colo.: New Science Library, 1984), 312-13.
19. See Virilio's contribution to this volume.
20. On the one hand, capitalist systems advocate stable values, the family, no abortion, and the like, while on the other destabilizing completely deprived segments of the population by exploding the familial cell and thereby taking away the last kernel of resistance. This can be seen in the alarming increase in the numbers of children living without parents.

The same goes for the concept of *atopia*, a destabilizing of the locus of power through a leftist critique and the kind of forced atopia—or teletopia—imposed on people away from friends and consuming in isolation what is imposed on them, living in the pure present with no memory or foresight. Drugs are also often imposed. It is less the junkie who is "bothersome" than the pusher who terrorizes. And in spite of a so-called war, drugs are sold in collusion with the government. Noriega was removed only after he became an economic threat to the government. (Note that this essay was written before "family values" became an embattled issue of the 1992 U.S. presidential campaign.)

21. Guattari, *Les trois écologies*.

22. Gilles Deleuze and Félix Guattari, "La géophilosophie," in *Qu'est-ce que la philosophie* (Paris: Minuit, 1991).

23. See Michel de Certeau, *L'invention du quotidien* (Paris: Bourgois, 1980). Next to Lévi-Strauss, de Certeau is possibly one of the most ecological thinkers writing today. I discuss their work in my forthcoming study, *Eco-Subjects*, to be published by Routledge in 1994.

24. For a psychoanalytic approach to this very question, see Teresa Brennan's contribution to this volume (chapter 6).

25. For the same reason, the inane debates that we see aired in North America about ecology *versus* human employment preconceive technology as masculine, aggressive, and salubrious. The binary terms in which the debate is cast reduce ecology to a level that, for strategic reasons, dispels critical thinking.

26. In a recent interview, Claude Lévi-Strauss declared that, for him, it is not the presence of American culture on French television that is to be deplored, but the dubbing, and hence the loss of tongues that, at the same time, he sees as inevitable. This is one of many symptoms of global impoverishment. See Susannah Patton, "Defending Our Rites," *Paris Passion* (1991), 31-35.

Chapter 6

Age of Paranoia

Teresa Brennan

Summing up the "doctrine of antiquity" in a sentence, Benjamin writes: "They alone shall possess the earth who live from the powers of the cosmos."[1] He continues:

> The exclusive emphasis on an optical connection to the universe, to which astronomy very quickly led, contained a portent of what was to come. The ancients' intercourse with the cosmos had been very different: the ecstatic trance. . . . It is the dangerous error of modern man to regard this experience as unimportant and unavoidable, and to consign it to the individual as the poetic experience of starry nights. It is not; its hour strikes again and again, and then neither nations nor generations can escape it, as was made terribly clear by the last war. . . . Human multitudes, gases, electrical forces were hurled into the open country, high-frequency currents coursed through the landscape, new constellations rose in the sky, aerial space and ocean depths thundered with propellers, and everywhere sacrificial shafts were dug in Mother Earth. This immense wooing of the cosmos was enacted for the first time on a planetary scale, that is, in the spirit of technology. But because the lust for profit sought satisfaction through it, technology betrayed man and turned the bridal bed into a bloodbath. The mastery of nature, so the imperialists teach, is the purpose of all technology. . . . [But] technology is not the mastery of nature but of the relation between nature and man. . . . In technology a *physis* is being organized through which mankind's contact with the cosmos takes a new and different form from that which it had in nations and families. One need recall

> only the velocities by virtue of which mankind is now preparing to embark on incalculable journeys into the interior of time.[2]

Benjamin wrote this in 1925-26, a year before Heidegger published the treatise that was to become *Being and Time*. Like Heidegger, he thinks about the relation between technology and the mastery of nature and between physics and metaphysics. Yet Benjamin ties technological mastery to capitalism and imperialism. He is also politically optimistic about the possible outcome of this energetic unleashing, seeing it as the source of the proletarian revolts that accompanied and followed the First World War. Nonetheless, he left the matter of the relation between the physical cosmos and technological mastery at an allusive level. It is one that remains to be thought through. What follows will contribute to that thinking through by focusing on an argument implicitly forefronted in some feminist writing on psychoanalysis. This is the idea that psychical fantasies can be both foundational,[3] yet inflated or curtailed by changing sociohistorical circumstances. Still, it will be some time before it is clear how this idea illuminates the relation between technological mastery and the "physis of the cosmos." It will be some time because the precise nature of the psychical fantasy at issue remains to be clarified. Moreover, it will become clearer not through concentrating on the fantasies described in existing psychoanalytic theories, but through considering the desires encapsulated in consumer goods, or commodities. I shall begin with the desire for instant gratification. The desire for instant gratification is realized in a proliferation of commodities whose common denominator may be nothing more than the desire itself. The vending machine that provides instantly upon the insertion of a coin, the fast-food establishment that promises no delay, the bank card that advertises itself as the one that does away with the need to stand in a queue, all promise the abolition of waiting time. Yet a little reflection shows that commodities cater to more than a desire for instant gratification. They are also marked by an attitude of appealing availability: the "I'm here for you" message signified by the trolley at the airport that asks you to "rent me," or the advertisement that once asked you to "fly me." These appealing items are akin to those that promise service, such as the credit card that delivers the object of desire to your door: "Pick up the phone; we come to you." More than the abolition of waiting time is offered here; one will also be waited upon. And if the promise of service appeals to a desire for domination and control, it has to be noted that the illusion of control is also provided by vending machines and their ilk. The consumer makes it happen; or rather, the consumer is catered to through the fantasy of making it happen with minimal effort, even none at all. In this connection, the car is an exemplary commodity. It gives mobility without much activity to a passive director. At the same time, of course, it pollutes the surrounding environment.

As I have indicated, I want to propose that the desires encapsulated in commodities reflect an underlying transhistorical psychical fantasy. In other words, I am proposing that we treat the commodity as an external expression of that fantasy. But immediately this proposition raises three problems. The first has been indicated already; it is that the desires encapsulated in commodities do not tally exactly with any existing account of a psychical fantasy. The second problem is that of demonstrating that the fantasy expressed in commodities is in fact foundational. This problem is exacerbated by a third, namely, the problem of why it is that a foundational fantasy is externally expressed in a form that is on the sociohistorical increase. For commodities, whether in the form of consumer goods or in the form of the technologies that underlie their production, are evidently increasing.

We are not entirely in a void when it comes to considering these problems. While there is no extant account that tallies precisely with the fantasy I am assuming commodities encapsulate, a synthetic reading of certain psychoanalytic theories will provide one. In addition, that synthetic reading coheres because it makes central the fantasies Klein describes about the mother's body. This is appropriate in another way, given that it is a feminist reading of Klein, as well as of Lacan, that raises the question of the relation between psychical fantasies and sociohistorical circumstances.[4] Finally, the notion that a psychical fantasy can have an increasingly macrocosmic expression is not new. Something very similar is maintained by Lacan, who writes that we are living under the sway of a paranoid social psychosis, in an ''ego's era'' that began more than four centuries ago.

The next section outlines a synthesis of psychoanalytic theories based on the desires I am assuming commodities encapsulate. Specifically, it draws on Freud and Klein. A brief third section sketches Lacan's theory of the paranoid ego's era. On this basis, I return to the relation between the psychical fantasy and its global enactment in a proliferation of instantly gratifying servile commodities, and speculate on the physics of the process involved.

Persistently, consumer goods appeal through visual media. This, together with the desire for instant gratification these commodities encapsulate, directs us to Freud's pleasure principle. Freud's pleasure principle, more strictly his principle of *Unlust*, or unpleasure, as he first defined it, is about a hallucinatory visual world where instant gratification is paramount. It is also about how psychical reality, as distinct from ''material reality,'' comes into being.[5] When the longed-for object (initially the breast or mother) is not present, it is hallucinated in its absence. This hallucination founds psychical reality; the breast is present in the imagination, but not present in the material here and now. The act of hallucination provides instant gratification, but the satisfaction it affords is only short-term. For the breast is longed for because the infant is hungry, and the hallucination cannot appease the unpleasure of the need for food. In other words, unpleasure is due to the tension of need. Any need (to eat, urinate, defecate, ejaculate) increases quantitatively, and pleasure is felt when the need is relieved. A hallucinated breast does not of itself relieve the need. Indeed, it ultimately leads to more unpleasure, in that it generates motor excitations it cannot dispel; the expected satisfaction that accompanies the hallucination gears the body up, but the energy amassed through this excitement cannot be relieved, any more than the original need itself.[6]

It should be clear that Freud's (un)pleasure principle is an economic or quantitative one: it is about the quantitative buildup of tension or need. In Freud's own terms, it is a matter of psychical economy, loosely based in Fechner's psychophysics.[7] The economic or quantitative physical aspects of Freud's theory of the pleasure principle are frequently criticized. Its descriptive aspects are more generally accepted; few commentators have problems with the notion of instant gratification, or with that of visual hallucination. But if one reconsiders the desires implicit in commodities, it is plain that although the pleasure principle accords with the desire for instant gratification they express and with their visual presentation in various media, it does not account for the other desires revealed in their design, namely, the desire to be waited upon; the desire to believe one is the source of agency who makes it happen; the desire to dominate and control the other who is active in providing, but whose activity is controlled by a relatively passive director; and the aggressive desire toward the other, if we take pollution as evidence of aggression.

The last-named desire evokes Klein. In her theory, the infant desires to spoil and poison the breast (and the mother) with its excrement.[8] As well as desiring to poison, the infant desires to devour and fragment the mother's body. ''Cutting up'' the mother's body is a recurrent theme in Klein's analyses of small children. She ties this cutting impulse to the drive for knowledge: the urge to get inside, grasp, and in this sense understand what is hidden, and in the process destroy it.[9]

For Klein, the desires to poison, devour, and dismember, and to know through dismembering, are prompted by two interrelated forces. The first is the strength of the death drive working within. The second is the envy of the creativeness embodied in the mother and mother's breast. While the death drive and envy motivate these fantasmatic attacks on the breast, they also lead to a fear of retaliation. The fear is that the aggressed-upon breast will respond in kind; this fear results in what Klein terms the ''paranoid-schizoid'' position. It is paranoid because the infant projects its own aggressive desires onto the other, and the retaliation it fears (being cut up, poisoned, devoured) mirrors its own desires. It is schizoid because this paranoid projection involves a splitting both of the ego and of the other. For to deal with its dependence on the breast as the source of life, and its simultaneous fantasy that the breast is out to get it, the infant splits: there is a ''good'' breast, and a ''bad'' one. Yet because the badness the infant fears originates within itself, the splitting of the other presupposes and perpetuates a splitting of the ego. The ego, by depositing its own aggressive desires in the other, impoverishes itself by the splitting and by the repression or "denial" that this entails. The ego can recover its full potential only by reclaiming what has been cast out. This reclamation, when it occurs, can lead to depression: the recognition that the erstwhile projected badness lies within. It may also lead to reparation: the attempt to repair the damage done in fantasy.[10]

Leaving that hopeful note aside, it is important to add that the extent of the splitting, and of the poisoning, devouring, dismembering fantasies that accompany splitting, is mediated by anxiety. For Klein, anxiety derives from the death drive working within. In the last analysis, she posits that the strength of the death drive, and envy, is innate. Moreover, Klein's account of the splitting process presupposes a fantasy that has no direct correspondence with reality (the breast is not *really* cut up, etc.). It is a psychical fantasy, and clearly not a consequence of the infant's actual social environment nor of social events. It is also important to note that the splitting of the good and bad breast is remarkably similar to the splitting of women into two types: mother and whore. It is this splitting that constitutes the fantasy of woman, which Lacan believes is essential to the (masculine) subject's securing his sense of identity. For Lacanians, this fantasy is also meant to be transhistorical, which is why the issue of transhistoricity appeared on the feminist agenda in the first place. But as my immediate concern is with the desires encapsulated in commodities, I do not intend to address here the evident ethnocentric and historical problems raised by this transhistorical claim (are women always split into two types?). Enough to say that the Kleinian account of the splitting into good and bad has the dubious advantage that it ascribes the phenomenon to any subject, masculine or feminine. There is also more warrant for assuming that this splitting and the desires that accompany it in Klein's account have more compass, as we will see.

Thus far, we have a theory that accounts for the desire to poison or, in commodity terms, the desire to pollute. We also have some elements of a theory that accounts for the desire to dominate and control (insofar as the desire to get inside, cut up, devour, and so on involves control and domination). It remains to tie this theory to the instant hallucinatory gratification embodied in the pleasure principle, and the desire to be waited upon from a passive but authoritative position. Here Klein's analysis of envy provides an indirect clue:

> Though superficially [envy] may manifest itself as a coverting of the prestige, wealth and power which others have attained its actual aim is creativeness. The capacity to give and preserve life is felt as the greatest gift, and therefore creativeness becomes the deepest cause for envy. The spoiling of creativity implied in envy is illustrated in Milton's *Paradise Lost*, where Satan, envious of God, decides to become the usurper of Heaven. Fallen, he and his other fallen angels build Hell as a rival to Heaven, and becomes the destructive force which attempts to destroy what God creates. This theological idea seems to come down from St.

> Augustine, who describes Life as a creative force opposed to Envy, a destructive force.[11]

This passage is interesting because it points out, although it does so obliquely, that envy superficially focuses on attributes or possessions, rather than on the creative force that may (or may not) result in them. The passage also points out that envy will attempt to rival that which it envies, and that it will do so by constructing an alternative. More generally, Klein's analysis of envy in the essay from which the above quotation comes shows that while envious motivations are readily recognizable in destructiveness or calumny, they are less recognizable, although present, in denial. This is the form of denial that simply ignores or forgets whatever is displeasing to the ego. It is present in the denial of the labor involved in creativity. We recognize it where creativity is seen as accidental, or where it is attributed to a circumstance or to a possession.

Let us add to these observations a notion that is best elaborated by Freud. This is that the infant, or small child, imagines, in a reversal of the actual state of affairs, that the mother is a dependent infant.[12] In reversing the passive experiences of childhood into active ones in his play with a cotton-reel, Freud's grandson not only masters the mother's absence and introduces himself to deathly repetition,[13] he simultaneously enters the world of language through the mother's absence, which forces him to call. He also makes the mother into a fantasied small child that he controls, a child that is also an inanimate thing. If the notion of the reversal of the original state of affairs is made central, rather than the incidental aside it is for Freud, it has the advantage that it reconciles otherwise diverse findings. When realities are seen in terms of their opposites, the fact of nurturance and the means to grow becomes a threat to narcissism; it establishes the reality of dependence. From this perspective, the envy of the mother's breast is the resentment of that dependence, and the reason that nurturance, or love, or protection, or assistance is interpreted as assertion of superiority and power. "Only saints are sufficiently detached from the deepest of the common passions to avoid the aggressive reactions to charity."[14] There is a related, if less relevant, offshoot of the reversal of the original state of affairs into its opposite, an offshoot that we might usefully term "imitating the original," in which rivalry with the original is clearly apparent. The child imitates the mother; the commodity, harking back to this essay's point of departure, is often an imitation of the original.[15] While writing this, I went to the corner store for orange juice, and found only artificial orange drink in an orange-shaped container (with green leaves). I also took in late-night television, worst among it *The Stepford Wives*, which is all about constructing a reliable and completely controllable imitation of the original wife and mother, and *Star Trek II*, in which Project Genesis shows us humans reinventing the entire process of creation.

But keeping to the main thread: the tendency to look at realities in terms of their opposites is manifest at another level, which explains the desire to be waited upon. Originally, the infant is perforce passive, and is dependent on the mother's activity for survival. Yet it would be consistent with a fantasmatic reversal of the original state of affairs if the infant were to correlate its actual dependent reality with the fantasy of control through imagining that the mother's activity takes place at its behest. The infant does not wait upon the mother; the mother waits upon it. It is precisely this fantasy that is catered to by the commodities with which we began. But a little of reality lingers on, in the association between passivity and luxury, which recognizes that it is not the passive controller, or ''the infant,'' who labors. At the same time, the labor or activity involved in fulfilling the wish is denied insofar as its intelligence is denied. In fantasy, the mental direction and design of what labor effects is appropriated, only the manual activity is left out. Thus the mental whim and control is the infant's. The work goes elsewhere.

The split occasioned by this fantasy prefigures a deeper dualism between mind and body, in which direction or agency is seen as mental and mindful, and activity, paradoxically, is viewed as something that lacks intelligence. By an ineluctable logic, the activity of women as mothers is presented as passive; in fantasy, it lacks a will of its own, it is directed. And because direction is too readily confused with a will of one's own, this denial can readily be extended to living nature overall. In this connection, it is worth noting that the oft-repeated association of women and nature can be explained not by what women and nature have in common, but by the similar fantasmatic denial imposed upon each of them. In the case of women, it is the will that is denied. In the case of living nature, its own inherent direction is disregarded. But this is to anticipate.

As I have indicated, the fact that creativeness is not viewed as intelligent or directed activity is consistent with envy's predilection to focus on it as the possession of certain attributes, rather than as a force in itself. Creativeness is seen less as what one does than as what one has. Or, to say a similar thing differently, the dialectics of envy conduct themselves at the level of images. What matters is the appearance of the thing, rather than the process of which it is part. To say that what is envied is the mother's possession of the breast is to work already within the terms of envy, which are those of possessions, things, appearances, discrete entities, separable and separate from an ongoing process. Which brings us to the crux of the matter. Although a fantasy of controlling the breast cannot survive at the level of feeling (pain or pleasure), it can survive at the literally imaginary level of hallucination. In fact, the controlling fantasy can be perpetuated through hallucinations, and this ability to perpetuate it must contribute to the addiction to the pleasure in hallucination, despite its unpleasure at other levels. In other words, by this account, the fantasy of controlling the breast and the act of hallucination are one and the same, which means that the amazing visual power of hallucination is tied to an omnipotent desire from the outset.

Of course, feelings of omnipotence, for Freud, are infantile in origin, and also tied to narcissism. There has been some discussion of how it is that narcissism can come into being only through fantasy or hallucination, but the other side of this issue, which is how it is that hallucination is by nature an omnipotent or narcissistic act, has not been discussed.[16] It is one thing to concentrate on how it is that the subject's sense of itself as a separate being is inextricably linked to narcissism; that is to say, that it is only by the narcissistic act of fantiasizing about its own body or circumference that it establishes its separate self. It is another to think about how the narcissism involved is also, and simultaneously, an omnipotent fantasy about controlling the other, for to establish itself as separate, the subject has to have something to be separate from. This much is foreshadowed by Lacan.[17] But, by this account, the thing the subject is separate from is the breast or mother it imagines as available to it, subject to it, and toward whom it feels the aggressive desires that lead in turn to paranoia. Moreover, in the omnipotent act of hallucinating a breast it controls, the nascent subject separates and gives priority to its own visual capacity for imagination over its other senses. It is this visual capacity that allows one to imagine that things are other than as they are; to focus on the distinctiveness of entities other than oneself, rather than on the senses or feelings that connect one with those others; to believe in (and even achieve) a situation in which mental design and direction can be divorced from bodily action.

It seems we have an account of a psychical fantasy that tallies with the desires encapsulated in commodities. It is this fantasy that I am positing as the foundational fantasy. That is to say, I am positing that the desire for instant gratification, the preference for visual and "object"-oriented thinking this entails, the desire to be waited upon, the envious desire to imitate the original, the desire to control the mother, and to devour, poison, and dismember her, and to obtain knowledge by this process, are part of an original human condition. There will be more argument on why these aggressive desires are part of the human condition, and the forces that prompt the anxiety and fear that underlie them, in the concluding section. The immediate question concerns how it is that the commodities in which these desires were first discerned appear to be proliferating, as, perhaps, are the desires themselves. For, as I stressed at the outset, it is one thing to say that a psychical fantasy is foundational. The sociohistorical force of that fantasy, in different times and places, is another matter altogether.

As noted above, the idea that a foundational fantasy can be played out with more or less sociohistorical force is implicit in Lacan's theory of a paranoid social psychosis and an ego's era. In this brief sketch of this neglected theory of Lacan's, it is useful to begin with a quotation in which Lacan, like Klein, also refers to Augustine, who, of course, wrote long before the ego's era began, which was, according to Lacan, in the late sixteenth century. In this quotation, Lacan is discussing the individual ego as such, and the death drive:

> The signs of the lasting damage this negative libido causes can be read in the face of a small child torn by the pangs of jealousy, where St. Augustine recognized original evil. "Myself have seen and known even a baby envious; it could not speak, yet it turned pale and looked bitterly on its foster brother."[18]

After quoting Augustine, Lacan moves swiftly on to Hegel's master/slave dialectic and the attempted destruction of the other consciousness that the dialectic foretells. In another context, Lacan makes it plain that that dialectic is the key to the "most formidable social hell" of the ego's paranoid era, in that the era is built on a destructive objectification of the other, together with a destructive objectification in knowledge.[19] The nature of the destructive objectification involved in the master/slave dialectic is left largely unspecified, although Lacan indicates that it means turning the other into a controllable thing.[20]

The need to control is part of the paranoia of the ego's era; it results from the subject's belief that the object, the objectified one, is out to get it, but this paranoia originates in the subject's own projected aggressive desires toward the other. Nonetheless, its paranoia makes the ego anxious, and its anxiety makes it want to control. The objectification of knowledge is also paranoid; it is knowledge based on a need for control. It is knowledge tied to a "positivist" worldview in which what is seen, or what can be tested or proved to exist, especially on the basis that it can be seen, is privileged. The objectification of knowledge helps construct a world in which only objects (or discrete entities?) are recognized, and they can be recognized only by subjects. In turn (I think) these subjects are affected, if not driven, by the objects they construct, although Lacan does not pursue this point. Lacan is more concerned with the objectification of knowledge as such; in this concern, he is at one with Heidegger, to whom Lacan frequently alludes, although Heidegger centralizes the objectification of nature as "standing reserve" and the technocratic drive for mastery over nature in a way that Lacan does not.[21]

In fact, generally, when Lacan comes to describe how it is that the ego's era came into being, he is, aside from one brief argument, not preoccupied with its social dynamics. But it is this consideration that preoccupies the present essay. My main concerns are with how it is that a foundational fantasy goes beyond the bounds of individual dreams and makes those dreams come true in the ego's era, and how it is that gradually the ego's era spatially encompasses the world at large, as Lacan believes it does. To deal with these concerns, the nature of objectification needs to be defined more precisely than it was by Lacan, or by Heidegger. The true nature of objectification is better grasped by Klein, in her analysis of the infantile desire to poison, fragment, and destroy the mother's body. By this argument, these desires constitute the process of objectification.

We have seen that turning the other into an object also means fragmenting it (in order partly to know it) or poisoning it or in other ways attacking it, as well as making it a controllable thing. A very similar point is made by Kristeva, who, in an argument that echoes that of Mary Douglas, makes "abjection" the foundation of objectification. Abjection is the feeling that one has of revolting (including excremental) substances within; objectification comes from the need to exclude these substances by depositing them in the other, which brings the other, as object, into being.[22]

Some of the resonances between Klein's theory of the infant and Lacan's theory of the ego's era should now be evident. I will assume that the links between their arguments on the role of anxiety in "objectification" can be taken for granted. Also, as Klein's account ties the objectifying desires to the drive for knowledge, it is not difficult to leap from it to Foucault's analysis of the drive for knowledge as a drive for power, a jump that is facilitated by the similarity, or indebtedness, of Foucault's theory to Heidegger's. Yet in making the leap between the transhistorical fantasy described in the preceding section and the processes at work in the ego's era, Klein's "mother body" has to be correlated with living nature. It is this correlation that is the condition of recognizing that the process Klein describes is a microcosm of a large-scale assault, a psychical fantasy writ large. This correlation is also necessary to begin answering the question, What is the relation between the psychical fantasy and sociohistorical processes that makes the dream of mastery over the mother/earth come true? Lacan refers to an ego's era, yet it is unclear how the historical era and the ontological ego enact the same desires and fantasies, one on a macroscale (as it were) and one on a microscale. As mentioned above, Lacan has only a brief argument concerning the ego's era. This argument, which concerns spatiality and the relation of spatial restrictions to psychical aggressiveness, will be relevant subsequently, but it will be so only after the general relevance of spatiotemporal considerations to the questions at issue has been established. Moreover, it is only then that it will be plain how the production of commodities, or consumer goods, in which the desires embodied in the foundational fantasy were first discerned, leads to the global technocratic expansion that marks the ego's era.

Evidently, as we have every reason for supposing that the fantasy of subjecting and dismembering that on which we depend is an ancient one, then its microcosmic version predates both the technocratic acting out of that fantasy on a large scale and the proliferation of consumer goods that satisfy that fantasy in everyday life. Or, more accurately, as the very idea of a foundational fantasy entails the assumption that it occurs in individuals, its genesis, although something everyone experiences in the West, is more individually contained. The limits on the extent to which it can be acted out are set by the available technology. It does not automatically become a corporate process.

But the question of the relation between the individual foundational fantasy and its sociohistorical parallel is complicated by two other things. The first is the emphasis on the psychical fantasy of woman, which has obscured the real transhistorical fantasy at issue. By the foregoing account, the ego comes into being and maintains itself partly through the fantasy that it either contains or in other ways controls the mother; this fantasy, as discussed, involves the reversal of the original state of affairs, together with the imitation of the original. When recognition of the other is unavoidable, the ego's first response is that it is not the dependent child. In patriarchal societies, the fantasied reversal of the original state is actualized in the relation between the sexes. Herein lies the importance for a man of the need to take care of the other, to be the breadwinner, a matter whose significance may lie in the distance between the extent to which he actually gives of himself and the extent to which he relies on the other's giving him an image of himself as giving, regardless of the reality of whether he gives or not. But more to the point: the truly patriarchal society is on the decline.[23] It is easy enough to see how in such societies, the psychical fantasy of woman could present itself as the necessary condition for containing and expressing the splitting into good and bad categories that figure in the foundational fantasy. But by this argument, the construction of sexual identity is not the origin of the foundational fantasy. It is rather that in patriarchal societies, far more ancient psychical conflict is played out in the arena of sexual identity. It is noteworthy that the shift from a genuinely patriarchal feudal society to a sexist capitalist one is also the shift from a society with a limited technology to one that is capable of satisfying the desires in the foundational fantasy with more precision.[24]

The second, related, complicating factor involved in understanding the relation between the individual foundational fantasy and its sociohistorical enactment is that the acting out of the fantasy on a large scale also takes place over a longer time scale. Yet it is precisely this complication that suggests how the relation between the two levels might be understood, for the key terms it introduces are those of time and space. The large-scale acting out represents the fantasy's spatial and temporal extension. Instead of the length of an individual's lifetime, or the years of a person's madness, the ego's era spans a few centuries. Instead of fantasies that are dreamed, there are technologies that make them come true, increasing their coverage of the earth's surface and corruption of its parameters in the process. It remains, then, to examine these connections and their physical implications at more length.

Time and space have already been implicated in the dynamics whereby the individual foundational psychical fantasy is generated. We want it *now*, and we want it to come to us. On the social scale, by inventing technologies that bring whatever it is we want to us, and that do so immediately, we are abolishing time and space. But paradoxically, as we have seen, this entails extending the fantasy in space and, for a reason yet to be determined, giving it more time to play itself out. The only way of resolving this paradox is to suppose that as we extend the fantasy in space, and make it immediately present, we simultaneously slow down time. In turn, this means supposing that the mechanism by which we make the fantasy present and extend its spatial coverage also congeals or slows down time. What is this mechanism and how is the paradox to be resolved?

This paradox is resolved in the case of infancy and the birth of psychical reality in this way: what prompts the hallucination is the desire that the longed-for object be present here and now. Yet if we examine Freud's account of hallucination, we find hallucination not only introduces instant gratification (in theory); in practice, it also introduces delay. In Freud's terms, the secondary process comes into being through an inhibition (*Hemmung*) of the primary process.[25] In the primary process, almost all things are possible; it is governed by the pleasure principle and marked by hallucinatory wish fulfillment, a lack of contradiction, the much-discussed mechanisms of condensation and displacement, and timelessness, among other things. The secondary process is governed by the reality principle. It is the locus of rational thought, directed motility, and planned action or agency. When it inhibits the primary process, it checks out or "reality tests" whether the image before it is a real perception or an imagined hallucination. In other words, it makes the psyche *pause* before it responds to the image it is offered. So, on the one hand, hallucination inaugurates a delay; on the other hand, I would suggest that hallucination is a response to a delay, on the grounds that the wish for instant gratification must be prompted by the experience of a gap between the perception of a need and its fulfillment.[26]

In the social case, the mechanisms by which we extend the fantasy are territorial imperialism and technology. Technology constructs the commodities that satisfy the fantasy of instant gratification and service, but how do these constructions simultaneously slow things down? For by the parallel presented here, the commodity takes the place of the hallucination, yet there is no need to distinguish whether the perception of a commodity is real or imagined. It exists. So how, then, can the existence of commodities demand delay, or an inhibition of the primary process, in the same way that a hallucination does?

Things might be clarified if, instead of concentrating initially on the parallel between the commodity and the hallucination, we ask what, in the sociohistorical macrocosm, parallels the primary process, which means concentrating for a little on the primary process as such. The nature of the primary process is one of the most taken-for-granted yet confused areas of Freud's theory. In addition to the characteristics already noted, the primary process consists of freely mobile energy, and there are reasons for thinking Freud identified it with the "movement of life" as such. At the same time, the primary process consists of the pathways in which energy is bound, a bondage that leads to repetition, and repetition, in turn, is the hallmark of the death drive. I have analyzed this confusion elsewhere;[27] let me simply reiterate here that the bound and repetitive pathways of the primary process come into being *through* repression. In fact, they come into being *through* primal repression, which Freud distinguished from repression proper, or secondary repression. Primal repression (of some idea or ideational event) establishes a nucleus that attracts subsequent "proper repressions" toward it.

A loose analogy connects Freud's distinction between primal repression and repression proper and a distinction drawn by Laplanche and Pontalis between primary and subsequent fantasies.[28] Given that primal repression seems to pertain to hallucination, it is appropriate to ask whether hallucinations and primal fantasies are the same thing. There is no obvious answer to this question. Nonetheless, a distinction needs to be drawn between hallucination and fantasy, in the everyday sense of that term. A hallucination appears to be present here and now, while a conscious fantasy is the contemplation of an event that is not occurring here and now. In addition, a hallucination is a picture on a larger scale than a daydream. If I close my eyes to daydream, the images I have seem smaller or more distanced than the images I have if I hallucinate, as everyone does in a nightmare. This difference or distance is a sensory affair; as Freud noted, hallucinations have the quality of sensory immediacy, but everyday fantasies or daydreams do not. Something else characterizes daydreams, and this something is not only memory, as daydreams encompass more than actual recollection. This something must be nothing less than the capacity for abstraction. Abstraction in any form is the removal of the subject's attention, or, for that matter, the subject's theory, from its referents in immediate felt experience. Where this distinction between hallucination and everyday or conscious fantasy leaves unconscious fantasy is unclear.

What I want to suggest now is that the act of repressing a hallucination is basic to establishing a sense of space-time (and, perhaps, to establishing the repressed unconscious), in that it establishes a still point of reference from which the nascent ego can get its bearings. Literally, its spatiotemporal bearings. This means that the sense of perspective is a construction, as may be the sense of passing time. The idea that the sense of perspective is a construction is attested to by the fact that when sight is recovered after blindness, the sense of perspective (distance and size) does not necessarily accord with the perception of others. It is often completely out of proportion with what we know as reality. The idea that the sense of passing time is also constructed is demanded by the theory that space-time is a continuum; time is measured in terms of space, and the interval between one event and another depends on the speed it takes to cover the distance between them, and speed, in turn, depends on the potential motion or energy of the body involved.[29] But if one looks more closely at what the initial repression of hallucination involves (the process by which the hallucination becomes unconscious), it is evident that something is happening to energy in the process, and also that "time" is measured relative to something other than the constructed space-time of which it is also part.

I have suggested that a hallucination is prompted by the delay between the perception of a need and its fulfillment, and noted that Freud (although he does not postulate an initial delay) argues that the secondary process comes into being through an inhibition of the primary process, which in turn amounts to a further delay. Postulating an initial delay between the perception of a need and its fulfillment as the condition of hallucination means postulating a prior state in which perception and need coincide, or in which the delay between the need and its fulfillment was shorter. The fact that there is an intrauterine state that is experienced before birth meets the requirements for this prior state. That is to say, if we suppose that *in utero* there is no experience of a delay between perception and need, or that any delay is shorter, the intrauterine state should constitute another pole against which the construction of space-time could be measured. This supposition has more substance if one considers what happens to psychical energy when it is bound. Freud's argument on this (elaborated in his Project) has led to a debate among psychoanalysts as to whether the bound pathways that come into being through distinguishing between hallucination and real perception are on the side of the Life or the Death drive.[30] The key opposed positions here, which I shall sketch only briefly, are represented by Laplanche and Lacan, respectively.

Laplanche disputes Lacan's location of the ego on the side of the death drive.[31] He does so on the basis of an interpretation of Freud's assumptions about physics (which, writes Laplanche, were outdated even at the close of the nineteenth century). The essence of Laplanche's argument is that, first, the ego is a kind of giant fantasy in itself. This much he has in common with Lacan. Laplanche bases his view of the ego as a fantasy in itself on the *Project*, where Freud posits the ego as a mass of cathected neuron pathways. Or, if we put this in terms of Freud's subsequent, less physiological vocabulary, the ego is a mass of pathways in which psychical energy is bound. It would be interesting to investigate how this bound mass tallies with Lacan's mirror image, especially given that the mirror stage and the nascent subject's mirror image are critical in its establishing its spatial sense, but that is beyond my scope here.

Keeping to the main thread: Laplanche also argues that Freud confuses the physical principle of inertia with the principle of constancy. The former is a state in which there is no motion, nothing. It is the desire to restore an earlier state of things in which the governing principle is rest. In Freud's 1920 formulation on the death drive, he termed it the Nirvana principle.[32] The principle of constancy is the desire to keep energy constant. For Freud, freely mobile energy will follow the path of least resistance, which is the path toward Nirvana. For Laplanche, while the ego is a giant fantasy, it is nonetheless a vital one, in that its bound pathways are the essential means for action against or toward what is necessary for sustaining life. There is no essential contradiction between its actions toward or away from life and the principle of constancy.

Where things get more complicated is that, as we have seen, the bound pathways are also tied to the death drive. This is because energy will flow along the paths that are familiar to it, and these paths might be completely inappropriate for dealing with a novel situation. The repression that brings both the pathways and the ego into being figures here; bound energy flows along pathways that are unconscious. In addition, there are two complications that reinforce the notion that the bound is on the side of the deathly. The first is that the ego is less likely to adapt and follow new pathways in a situation that arouses anxiety. Furthermore, it is precisely the protracted attachment to any fantasy (which must necessitate a bound pathway) that characterizes neurosis. Such attachments make it harder to act upon the world; they are similar in their effects to anxiety, in that they counter "the movement of life."[33] Hence Lacan's position.

Now it would be easy enough to take a liberal approach here, and say that, on the one hand, the ego and bound pathways are necessary: one has to deal with life's exigencies (Laplanche). On the other, if too much psychical energy is bound, if the pathways are too rigid, if anxiety is greater, then vaulting ambition overleaps itself and the result is deathly (Lacan). But this balanced solution allows one supposition to escape unchallenged. This is the notion that as freely mobile energy follows the path of least resistance, it *therefore* tends towards inertia. There is a related supposition, which is that an inert state is a restful one, and that any body seeking rest will seek to be inert or motionless. There is actually no reason rest should be equated with inertia. As we have seen, the natural state (experienced *in utero*) could well be one of more rapid motion, and this living state could be restful in that it appears to be without the conflict contingent on delay, and is therefore "timeless." In other words, what leads freely mobile energy on its quest for the path of least resistance is not the notion of inertia, but the memory of a state of timeless (yet, relative to the subsequent sense of time, more rapid) motion. It is no accident that Aristotle, who argued that, although motion and time are mutually defined, motion depends on an unmoved mover, an ultimate still point, also argued that the mother's role in gestation is entirely passive. The metaphysical presupposition hidden behind the founding assumption of classical Aristotelian physics is consistent with Aristotle's denial of maternal activity. To suppose, as I have, that *in utero* there is no delay or less delay between a perception of a need and its fulfillment is to suppose that there is a system of fleshly communication between two parties. It means that the mother is not, as Aristotle has it, a passive garden in which a tiny, active, fully formed homunculus is planted, and grows of its own accord.[34]

Of course, the spatiotemporal notion of rapidity comes into being only after the fact, that is to say, after birth, and the experience of delay. The point is that, after the fact, the resultant slow plight of the ego is measured retroactively, in spatiotemporal terms, against the prior intrauterine state. In addition, the very thing that leads freely mobile energy into conflict with the exigencies of life is the fact that it encounters a point of resistance. If there is no resistance, there is no a priori reason why freely mobile energy could not regain its prior rapid motion. Naturally, this means external as well as internal points of resistance, for it would be a travesty of what logic underlies Freud's reasoning on the ego to reduce the points of resistance the ego encounters to its own self-sustaining fantasies. The ego evidently encounters other points of resistance that would harm its chances of living (very bad weather, aggressive others, and more), and to these it has to respond. Nonetheless, the notion that the ego's own hallucinated responses constitute the first point of resistance will be instructive if pursued in relation to the parallel drawn throughout this essay between the desires encapsulated in commodities and those of the psyche.

Let us suppose that the construction of a commodity also binds energy in the same way that it is bound in the repression of a hallucination. That is to say, the energy is attached to an image, fixing it in place. The energy attached in this way is that of living nature; it correlates with Freud's freely mobile energy, although freely mobile energy knows no paths except those that tend to Nirvana. Living nature, on the other hand, has its own rhythms and paths. Yet, as we have seen, the notion that the paths of freely mobile energy and those of the life drive are not the same depends on the idea that freely mobile energy follows the path of least resistance, and this in turn depends on the existence of something that resists. The commodity provides that point of resistance, in that it encapsulates living nature in forms that remove it from the flow of life. A tree converted into a table does not enter into the production of more trees. Such conversions, which of course become more significant when the commodity produced is not biodegradable, function analogously with fantasies in that they bind living substances in forms that are inert, relative to the energetic movement of life. The more of these relatively inert points there are, the slower the movement of life becomes. That is consistent with the phenomenon known as entropy. The implication here is not that nothing should be constructed (shelter etc.), any more than it means the ego should not respond to bad weather by protecting itself (seeking shelter or other acts).

However, the notion that points of resistance slow things down provides a critical principle by which to gauge what should and should not be constructed. That gauge is, how readily can these constructions reenter the movement of life? It also means we have an account of the paradox whereby the infantile fantasy takes more time to play itself out, and it is an account consistent with the idea that the fantasy simultaneously extends itself in space. It takes more "time" to play itself out in the sense that it uses more living energy, as it systematically extends itself in the spatial conquests necessary to supply the living substances by which it sustains itself. However, it also follows from this argument that the "time" it takes to play itself out is itself a constructed phenomenon, in that this "time" consists of the accumulation of "points of resistance" or commodities. Moreover, this "time" has its own direction. The construction of one commodity (using the term in the broadest sense)[35] fixes a relatively inert or still point. This point (let us say it is a factory, even a town) then functions as an inert point of reference from which distances are measured and pathways built. They are built, at least in part, as a means to further the consumption of more living substances in the process of production. Of the different characteristics that mark the networks established by these means, there are two that need to be noted here. The first is that to stay in the race of efficient consumption for production and further consumption, and free-enterprise competition is always a race, these networks need to facilitate the most rapid transport of energy possible. This applies to energy of any order: the natural substances consumed in production and human labor power. The means by which natural substance or labor is extracted and conveyed from *a* to *b* has to be speeded up. So at the level of constructed space-time, everything seems to be getting faster and faster. The second point about the networks constructed in relation to still points is that at the same time as they partake in the process whereby natural reproduction is actually slowed down, they must, like the still points themselves, have their own physical energetic effects.[36] These effects are physical in the sense that commodities function as points of resistance to natural rhythms, so that in reality things get slower and slower. The second point is that as the networks between these points extend, creating more still points in the process, the expanding spatiotemporal construction that results has a pattern of its own.[37]

There is every reason for supposing that this pattern presents itself to us as temporal causality. Temporal causality is the process whereby one thing appears to lead to another across time in an apparently irreversible manner. This taken-for-granted process is of course at issue in physics, where it is regarded as something to be explained. This analysis might, incidentally, contribute to an understanding of the asymmetrical nature of time—the puzzle as to why time goes only one way, or why time is irreversible. By this account time could be understood, in theory if not in practice, as reversible, provided that all the points of resistance out of which space-time is constructed and connected were systematically undone, and if their component natural substances reentered the natural rhythms of production from which they were initially, physically, "abstracted." This understanding of time also accords with the deconstructionist idea that causality is a construction, a line of reasoning we impose on events. Except that, in this case, the causal construction really has been constructed. The fact that the construction has a fantasmatic origin makes it no less physical in its effects. In other words, reading causality as a mere illusion that could be done away with by refusing to impose causal reasoning in theory accords with and therefore does nothing to counter the galloping construction of causality in the physical world.

This returns us to the question of how the infantile fantasy plays itself out on the larger scale over longer time. The dynamics described in this process must be cumulative, not only in the sense that as things get faster and faster at the level of constructed space-time, they get slower and slower in terms of the natural rhythms of reproduction. They must also be cumulative in terms of the extent to which the causality constructed presents itself to us as a historical process. "History," as the sense of the sequence of past events, is increasingly molded by the extent to which a foundational psychical fantasy makes itself materially true, and by its consequent material effects on the individual psyches that entertain the fantasy. If these material effects are taken into account, the extent to which the fantasy takes hold individually, and thus the extent to which individuals act in accord with the fantasy's constructed causal direction, might also be cumulative. At this point I return to Lacan.

When he discussed the ego's era, Lacan noted that it was accompanied by increasing aggressiveness and anxiety. He explained this in terms of the spatial constrictions of the urban environment. The more spatially constricted the environment becomes, the greater the anxiety and the greater the tendency to project this anxiety outward in aggressiveness. Lacan's account of this process is as allusive as his account of objectification. Here again, a link needs to be made between aggressiveness and the other dynamics of objectification revealed by the analysis of the commodity. At one level the link is self-evident. If a vending machine fails to produce, its chances of being kicked are high. At another level, it is plain that the link has to be established in spatiotemporal, physical terms. Lacan's emphasis on spatiality provides a physical pointer, which can be extended in terms of the speculative account of still points of resistance offered here. These still points are inanimate, whether they are constructed commodities or internal fantasies. What I want to suggest in concluding is that just as its own fantasies weigh heavily upon the ego, so does a subjective if subliminal sensing of what is animate or inanimate in the surrounding environment. The less animate that environment is, the slower time becomes, then the greater the ego's need to speed things up, its anxiety, its splitting, its need for control, its "cutting up" in its urge to know, its spoiling of living nature, and its general aggression toward the other. But of course, as with any paranoid anxiety, the ego, by these processes, only accelerates the production of the conditions that produce its fears.[38] It constructs more still points that start, or speed up, the whole show again.

The result is an increasing incapacity to tolerate delay, a greater demand for service, a more extensive need for domination, a horror of inferiority contingent on escalating envy and the constant comparisons envy demands, and an ever-rising flight away from the active living flesh into the fantasmatic world of metaphor. It is this last that makes the sometimes obvious nature of the processes recorded here elusive.[39] The originating foundational fantasy situates the mother as a passive natural entity responding to an active agency located elsewhere. The extent to which active agency really is located elsewhere increases as the material means to control the environment increase. In other words, in that active agency is the ability to do things according to one's own direction, to impose direction, to "make it happen," this ability must increase as the material means for accomplishing one's will also increase. To the extent that this active agency results in the imposition of a direction on the environment that goes against, rather than with, natural rhythms and their own logic, the force of the latter figures less in any calculations made about what causes what. To say the same thing differently, by this argument, the subject's sense of connection with the world is physically altered by its physical environment. And if the physical points of resistance embodied in commodities function after the manner of fantasies, closing the subject off to the movement of life, they are also visual tangible evidence of a different physical world that, however fantasmatic its origin, makes the subject more likely to see what it has made, rather than feel itself to be connected with, or part of, what has made it. The visual hallucination that denied feelings of unpleasure is now a concrete thing, and the various senses that otherwise connect the subject with the world stand back in favor of the visual sense. This visual favoritism takes us back to the optical connection with which we began.[40]

The idea that the subject's sense of perception is physically altered by its physical environment, and the correlative idea that the concrete imposition of a foundational psychical fantasy has altered that environment, also raises the possibility that different physical theories and theories of perception are more true for their times than they appear with hindsight, precisely because the times physically alter what and how we perceive.[41] If the parallel drawn here between psychical and sociohistorical temporal interference is correct, if the construction of more and more commodities slows down real time while seeming to speed it up, then this means that the physical reality in which we exist and the physical laws under which we live *are being and have been altered*. By a sociohistorical process, have we produced the chaotic *physis* we now discover, as if it had always been present? For if we have, we have done so by enacting a psychical fantasy that, because it relies on a divorce of mental and physical activity, reinforces the prejudice that the psychical process and its sociohistorical parallel have no effect on the physical world; this prejudice in thought may be why it is difficult to get clear information from a scientist concerning the question just posed.

Pending that information, there is no reason the process described here cannot be reversed or, at least, reversed to the extent that an awareness of the function of still or inert points of resistance means that their worst effects are avoided. In other words, we confront teleology only in the logical sense that the process described here is cumulative. Where the energy to reverse that accumulation comes from is another question, except that it comes. It is evident all around us. Perhaps Benjamin was right, and the very act of unleashing uncharted forces provides a fuel, or is it an intensity of will that can reverse direction and revivify what it has harmed? The owl of Minerva is beginning to flutter. She might yet fly, if she can free her wings of the oil slick.

Notes

1. Walter Benjamin, *One Way Street and Other Writings*, trans. E. Jephcott and K. Shorter (London: NLB, 1979), 103.

2. Ibid., 103-4.

3. Initially, I termed this psychical fantasy ''transhistorical.'' ''Foundational'' is a more accurate term. For an extended discussion of this substitution, see Teresa Brennan, *History after Lacan* (London: Routledge, 1993).

4. I elaborate this idea in Teresa Brennan, ''Controversial Discussions and Feminist Debate,'' in *Freud in Exile*, ed. Edward Timms and Naomi Segal (London: Yale University Press, 1988), 54-74. It is also referred to in the context of an extended discussion of Lacan's theory of history in the book from which this essay is extracted: Teresa Brennan, *History after Lacan*. For another discussion that refers to the difference between psychical fantasy and social reality, see Jacqueline Rose, ''Introduction II,'' in *Feminine Sexuality: Jacques Lacan and the Ecole Freudienne*, ed. Juliet Mitchell and Jacqueline Rose (London: Macmillan, 1982), 27-57.

5. The distinction between psychical and material reality is Freud's. It has been criticized by J. Laplanche and J.-B. Pontalis in ''Fantasy and the Origins of Sexuality,'' *International Journal of Psychoanalysis* 49 (1968): 1-18. Their criticism is made with good reason, as will be plain in what follows.

6. See in particular the well-known seventh chapter of Freud's *The Interpretation of Dreams*, vol. 5 of *The Standard Edition of the Complete Psychological Works*, ed. James Strachey, trans. James Strachey et al. 24 vols. (London: Hogarth Press and the Institute of Psychoanalysis, 1953-74). (hereafter *SE*).

7. For the most thorough discussion of Freud's relation to Fechner, see H. F. Ellenberger, *The Discovery of the Unconscious: The History and Evolution of Dynamic Psychiatry* (London: Allan Lane, 1970).

8. In discussing the infant's desires in Klein's theory, I should enter a brief caveat on the notion that ''the infant'' is the sole culprit when it comes to pinpointing the origin of the aggressive desires under discussion. ''The infant'' is always that origin for Klein, although we will see later that the question of culpability is more complicated. For the time being, however, I shall continue to write in terms of the infant.

9. For representative illustrations of these and many of the following Kleinian ideas from different periods of Klein's work, see Melanie Klein, ''Early Stages of the Oedipus Conflict,'' in *Love, Guilt and Reparation and Other Works, 1921-1945: The Writings of Melanie Klein*, vol. 1 (London: Hogarth, 1985), 186-98; and ''Envy and Gratitude,'' in *Envy and Gratitude and Other Works, 1946-1963: The Writings of Melanie Klein*, vol. 3 (London: Hogarth, 1980), 176-235.

10. The most representative if difficult account of the views summarized in this paragraph is found in Melanie Klein, ''Notes on Some Schizoid Mechanisms,'' in *Love, Guilt and Reparation and Other Works, 1921-1945: The Writings of Melanie Klein*, vol. 1 (London: Hogarth, 1985), 1-24.

11. Klein, ''Envy and Gratitude,'' 201-2.

12. Significantly, Freud mentions this in his discussion on ''female sexuality''; *SE*, vol. 21, 236.

13. Sigmund Freud, *Beyond the Pleasure Principle*, in *SE*, vol. 18.

14. Jacques Lacan, ''Aggressivity in Psychoanalysis,'' in *Ecrits: A Selection*, trans. Alan Sheridan (London: Tavistock, 1977), 13.

15. The imitation of the original is an often implicit and sometimes explicit theme in discussions of women and technology, particularly reproductive technology. For a general representative collection on this theme, see Michelle Stanworth, ed., *Reproductive Technologies* (Cambridge: Polity, 1987). For discussions that bear more closely on the issues discussed here, see Donna Haraway, ''A Manifesto for Cyborgs: Science, Technology, and Socialist Feminism in the 1980's,'' *Socialist Re-*

view 80 (1985): 65-107; Rosi Braidotti, ''Organs without Bodies,'' *Differences* 1, no. 1 (1989): 147-61.

16. However, Borch-Jacobsen comes close when he pinpoints the core of megalomania in many of the dreams Freud analyzed; Mikael Borch-Jacobsen, *The Freudian Subject* (London: Macmillan, 1989). Borch-Jacobsen's analysis of why narcissism is necessary, and is in fact the key to the constitution of the subject, is the outstanding discussion of this theme. Laplanche and Pontalis's classic ''Fantasy and the Origins of Sexuality'' is also important, as is J. Laplanche, *Life and Death in Psychoanalysis*, trans. J. Mehlman (Baltimore: Johns Hopkins University Press, 1976).

17. It is the essence of Lacan's concept of the *objet petit a*.

18. Jacques Lacan, ''Some reflections on the ego,'' *International Journal of Psychoanalysis* 34 (1953): 16. Lacan does not give a reference for the quotation from St. Augustine. It comes from the *Confessions*. The context is an argument that sin is present in infancy. When considering various possibilities, St. Augustine asks whether as an infant he sinned by endeavoring to harm ''as much as possible'' those larger beings, including his parents, who were not subject to him, ''whenever they did not punctually obey [his] will [*non ad nutum voluntatis obtemperantibus feriendo nocere niti quantum potest*].'' One sentence later comes the observation that Lacan also quotes, in part, in Latin: ''*vidi ego et expertus sum zelantem parvulum: nondum loquebatur et intue batur pallidus amaro aspectu conlactaneum suum*.'' Aureli Augustini, *Confessionum*, ed. P. Knoll (Leipzig: Verlag von B. G. Teubner, 1898), 1/14-20, p. 8. For my purposes, it is the failure to obey *punctually*, in connection with envy, that is interesting.

19. Lacan, ''Aggressivity in Psychoanalysis,'' 79. This and ''The Function and Field of Speech and Language in Psychoanalysis,'' in *Ecrits: A Selection*, trans. Alan Sheridan (London: Tavistock, 1977), 30-113, are the main texts in which Lacan outlines his theory of the ego's era.

20. Lacan, ''The Function and Field of Speech,'' 42.

21. Martin Heidegger, *The Question Concerning Technology and Other Essays*, trans. W. Lovitt (New York: Harper and Row, 1977).

22. J. Kristeva, ''L'abjet d'amour,'' *Tel Quel* 91 (1982), 17-32; Mary Douglas, *Purity and Danger: An Analysis of Concepts of Pollution and Taboo* (London: Routledge and Kegan Paul, 1966). Douglas's cross-cultural inquiry lends further weight to the notion that what we are dealing with here is a transhistorical fantasy.

23. See Juliet MacCannell, *The Regime of the Brother* (London: Routledge, 1991).

24. It is understandable that Lacan should locate the psychical fantasy of woman as the significant one. The odd thing is that he pinpoints it in terms of feudal societies, yet he nonetheless sees the subject/object distinction as a sexual fantasy that has governed thought since the pre-Socratics. And when he describes the subject/object distinction as a sexual fantasy, Lacan attributes the tendency to split form from matter, and to see matter as passive, to it. See Jacques Lacan, *Encore: Le séminaire XX*, trans. Jacques-Alain Miller (Paris: Seuil, 1975), 76. This accords with the above argument on the fantasmatic denial of intelligence in the bodily activity of women as mothers.

25. Freud, *The Interpretation of Dreams*, 601.

26. Of course, this has consequences for real perception as well as imagined hallucination. One cannot respond immediately to the former. It has to be evaluated. Nothing visual can be taken for granted. What is more, it only becomes taken for granted at the price of establishing familiar pathways for psychical energy to follow. The more certain the pathway, the less energy is involved; the less the pathways are disrupted, the less the stress. Yet the more the pathways are fixed, the more energy, in Freud's terms, is bound: rigidity and anxiety, in the face of unfamiliar pathways, are the consequences. Two points are critical here. The first is Freud's notion that energy comes to exist in a bound, rather than freely mobile, state through checking out hallucinations. The second is that, as we will see, Lacan ties anxiety to spatial constriction. The argument on bound and freely mobile energy is one that Freud attributes to Breuer and Breuer attributes to Freud. See Sigmund Freud, *Studies on Hysteria*, in *SE*, vol. 2, 194, n. 1.

27. I have also elaborated on the significance of *delay* in hallucination between the perception of a need and its fulfillment. See Teresa Brennan, *The Interpretation of the Flesh*, vol. 1, *Freud's Theory of Femininity* (London: Routledge, 1991), ch. 6.

28. Laplanche and Pontalis, "Fantasy and the Origins of Sexuality."

29. There are several excellent histories of physics and the story of the shift from one dominant paradigm to another, but see in particular Jennifer Trusted, *Physics and Metaphysics* (London: Routledge, 1991). For reasons of space, I cannot give an exposition of the relevant paradigm shifts here.

30. Freud, *Project for a Scientific Psychology* [*Entwurf einer Psychologie*], in *SE*, vol. 1; see also note 25 above.

31. Laplanche, *Life and Death in Psychoanalysis*.

32. Freud, *Beyond the Pleasure Principle*, 56.

33. Freud, *Inhibitions, Symptoms and Anxiety*, in *SE*, vol. 20, 148.

34. Aristotle, "Generation of Animals," in *Physics*, vol. 4, trans. Philip Wicksteed and Francis Cornford (London: Loeb/Heinemann, 1979), 120-22. On time and motion, see *Physics*, IV, XI, 219-20, pp. 389-95. It should be added that the confusion between "at rest" and "inert" is a subsequent one. Aristotle distinguishes the two states (*Physics*, 5.6.230), but immediately argues that "at rest" is analogous to unchangingness.

35. A commodity varies in scale depending on the position of the consumer. From the standpoint of one consumer, it could be a small-scale consumer good. From the standpoint of another, it could be ICI. Marx's definition of the commodity, which haunts this essay, should be discussed here, but lack of space prevents it.

36. The notion that instruments from "the microscope to the radio-television" have a far greater effect on us than we are aware has been put forward by Lacan, *Encore*, 76, and a similar observation is made by Benjamin, as noted at the outset.

37. At this point I must note the argument of Niklas Luhmann, who reiterates the Parsonian point that a temporal social system can be compared only with something that is not temporal; Niklas Luhmann, *The Differentiation of Society* (New York: Columbia University Press, 1982), 292. He even says that it can be compared only with something that is "immediate," that is to say, timeless. But having had this insight, he then goes on to neglect its implications altogether. He forgets the existence of the "immediate" something, and argues instead that a temporal system can have nothing outside itself, and thus that there is no point against which an alternative future to the one already contained within the present can be built. From the perspective of this argument, of course, that alternative point is present in the natural world. Moreover, as I imply in the text, if the "immediate" something is the physical world, if the physical world is also spatiotemporal, from where does it get its temporality, if not from the social world? For an excellent critique of Luhmann's "privileging of the present" from a Derridean perspective, see Drucilla Cornell, "Time, Deconstruction, and the Challenge to Legal Positivism: The Call for Judicial Responsibility," *Yale Journal of Law and the Humanities* 2 (Summer 1990): 67-97; Cornell's argument prompted this note. The significance of time and space in social organizations generally, and the shift from modernism to "postmodernism" in particular, is the central theme of Anthony Giddens, *The Consequences of Modernity* (Stanford, Calif.: Stanford University Press, 1990).

38. There is a striking account of the paranoid's collusion in the production of the conditions it fears in Bersani's discussion of Thomas Pynchon. See Leo Bersani, *The Culture of Redemption* (Cambridge, Mass.: Harvard University Press, 1990), 188.

39. What confuses the issue further, I suspect, is that the accumulating physical pressure is then deflected on to women, which means both that the tendency literally to objectify women at every level should be increasing and that one is likely to mistake the object onto, or from which, pressure is deflected as its cause. Thus, perhaps, the ludicrous projections that men direct toward women, claiming that women are really in control. On the other hand, as we have seen, the objectification of

women is not, in the last analysis, a product of the psychical fantasy of woman, but of a more ancient transhistorical fantasy.

40. The critique of ocularcentrism is so basic to twentieth-century French critical theory that extended references seem superfluous. The key figures here, of course, are Merleau-Ponty and Foucault. There is a useful critique and review of the relevant literature in M. Jay, "In the Empire of the Gaze: Foucault and the Denigration of Vision in Twentieth-Century French Thought," in *Foucault: A Critical Reader*, ed. D. Couzens Hoy (Oxford: Basil Blackwell, 1986). Rosi Braidotti develops the critique of ocularcentrism from a feminist perspective in "Organs without Bodies."

41. Luce Irigaray argues in *Ethique de la difference sexuelle* (Paris: Minuit, 1984) that any epochal change also requires a change in how space and time are perceived. The difference between her observation and this argument is that Irigaray sees the change in space-time perception solely in subjective terms, rather than attempting to explain it as a change that is physically produced. For a good discussion of French feminist theory and spatiotemporal concerns, see Elizabeth Grosz, "Space, Time and Bodies," *On the Beach*, Sydney, 13, 1988. It should be added here that many feminist philosophers of science have suggested that the boundaries of the physical sciences and their underlying assumptions should be thrown into question; there is no reason the physical sciences should escape attempts at gender blindness when the social sciences have not. But there is a difference between noting that the physical sciences should be questioned, or even showing that a more empathic approach to science is possible (as does Evelyn Fox Keller, for instance, in "Feminism and Science," *Signs* [Spring 1982]: 589-607) and proposing an alternative, speculative, physical theory. This is not the exhortation. This is the act.

Chapter 7

Heidegger and the Rhetoric of Submission: Technology and Passivity

Ingrid Scheibler

> Chronologically speaking, modern physical science begins in the seventeenth century. In contrast, machine-power technology develops only in the second half of the eighteenth century. But modern technology, which for chronological reckoning is the later, is, from the point of view of the essence holding sway within it, the historically earlier.[1]

With this well-known formulation, Heidegger distinguishes technology in its various concrete manifestations from what he calls its essence. In contrast to the changing developments in actual machine technology, the essence of technology—though it too is historical—is something fixed. This essence is not itself technological: Heidegger describes it as a bringing forth, which is grounded in revealing (*aletheia*).[2] Heidegger returns to the Greek *techné*—the activities and skills of the craftsman or artisan—to bring out its relation to the "bringing forth" of *poiesis*. In contrast to the Greek relation of *techne* to *poiesis*/bringing forth, modern technology does not engender a bringing forth in the sense of *poiesis*: "The essence of modern technology shows itself in what we call Enframing. . . . It is the way in which the real reveals itself as standing-reserve."[3] *Modern* technology, Heidegger says, is linked to a particular mode of conceiving our relation to the world—of bringing forth—through a process that objectifies that world. Moreover, the way in which everything shows itself (the unconcealment) at any given time may be misinterpreted in the modern period merely and wholly as standing-reserve. Heidegger points to the danger of this historical moment:

> As soon as what is unconcealed no longer concerns man even as object, but does so, rather, exclusively as standing-reserve, and man in the midst of objectlessness is nothing but the orderer of the standing-reserve, then he comes to the very brink of a precipitous fall; that is he comes to the point where he himself will have to be taken as standing-reserve. Meanwhile man, precisely as the one so threatened, exalts himself to the posture of lord of the earth. In this way *the impression comes to prevail that everything man encounters exists only insofar as it is his construct.*[4]

In Heidegger's analysis, to understand the essence of technology is part of a broader project of understanding the relation of this mode of objectifying experience to the tradition of Western metaphysics. In this sense, Heidegger links the historical fact of developing technologies to something that is historically prior to the specific sociocultural context of those technologies (reproductive technologies, weapons system technologies, genetic engineering, and virtual reality are some current examples). For Heidegger, then, this means that the question concerning technology cannot be thought apart from the critique of Western metaphysics.

Heidegger provides a rich resource for understanding—and therefore for assessing—our current global situation of environmental degradation, predicated as it is on what he describes as man's conception of the surrounding world as his own construction and manipulable resource.[5] Heidegger's writings on technology are inextricably bound to his broader critique of Western metaphysics and probing of the question of Being. They therefore concern the specific characteristics of modern technology and the attendant rise to dominance of modern subjectivism (the philosophy of consciousness) and its correlate—a conception of the real based on an ethos of mastery and domination that Heidegger describes as the modern relation to the real in terms of its objectification. Heidegger writes, "The revealing that rules in modern technology is a challenging [*Herausfordern*], which puts to nature the unreasonable demand that it supply energy that can be extracted and stored as such."[6] It is in conceptualizing an alternative to this modern ethos (of objectification, domination) that Heidegger's explorations are valuable, and I wish in this essay to emphasize their relevance to rethinking the privileged place of subjectivity and the sanctioning of its own ceaseless *activity* that Heidegger describes as part of the essence of *modern* technology.[7] In seeking an alternative conceptualization, I will be examining Heidegger's account of two types of passivity: the traditional-metaphysical conception of the relation between activity and passivity, and his distinction from this of another mode of relation, which he often articulates through a rhetoric of submission, that is, a call to be passive and yielding. It has been greatly underestimated that Heidegger does in fact rely on such a distinction. The latter relation, Heidegger says, is a passivity that lies beyond the traditional distinction between activity and passivity. To characterize this traditional distinction briefly, it presupposes that to be active is to act upon (something external) and to be passive is to be acted upon (by something external). Both activity and passivity as traditionally conceived presuppose a separation of subject and object, and in this way the privileged autonomy of subjectivity is preserved. That is to say, activity is the privileged term, for passivity is seen to presuppose a threat to the autonomy of subjectivity, that is, its subjugation or weakness of will. Heidegger's alternative account of passivity is explicitly articulated in his "Conversation on a Country Path about Thinking"[8] and is often expressed more generally (and less explicitly) in the rhetoric of submission used throughout his later writings.

My reading of Heidegger here will be a qualified defense, and this should be clear from the discussion itself.[9] This discussion was originally conceived as a response to Jürgen Habermas's reading of Heidegger in his lecture, "The Undermining of Western Rationalism: Martin Heidegger."[10] I wish to emphasize that my approach was conceived specifically as a response to Habermas's critique of Heidegger.[11] I will outline only the substance of Habermas's criticism before turning to my examination of Heidegger. In the first section of what follows, I will focus on Habermas in a summary that is not *directly* related to technology. I need to do this, however, in order then to draw out Heidegger's understanding of passivity, which is relevant to technology.[12] I then offer a qualified defense of Heidegger. Here, I begin by briefly characterizing Heidegger's project and, in the qualified defense, return to Heidegger's analysis of technology as such, linking it to the notion of meditative thinking put forward in his "Conversation on a Country Path about Thinking."

Habermas's Criticism of Heidegger

In his recent work, Habermas has devoted his attention to examining and defending what he calls the "philosophical discourse of modernity." In his lecture on Heidegger, Habermas takes issue with the hermeneutical-phenomenological project by evaluating Heidegger as an antimodern thinker. In his appraisal of Heidegger, Habermas raises several issues that lie at the heart of the debate concerning the nature of modernity:

1. examining the possibility of a critique of modern subjectivism, of transcending the philosophy of the subject;
2. assessing the attitudes to and conceptions of the modern Enlightenment project, its concept of reason; and
3. evaluating the critical potential of such a project—particularly the need to reconceive our relation to both the natural and social worlds in a postmetaphysical project critical of the dominating force of instrumental reason.

Habermas links Heidegger's later writings, which he also characterizes with reference to Heidegger's ideas of subjection and submission, to Heidegger's involvement with the National Socialist movement. He writes:

> I suspect that Heidegger could find his way to the temporalized *Ursprungsphilosophie* of the later period only by way of his temporary identification with the National Socialist movement—to whose inner truth and greatness he still attested in 1935.[13]

Habermas's discussion of Heidegger's political embrace of National Socialism raises an extremely important question that should be addressed in any assessment of Heidegger's work: the issue of how the facts of Heidegger's *political* involvement should affect an evaluation of his *philosophical* work. Whereas Nietzsche's work was appropriated posthumously, Heidegger's involvement was active, as exemplified by his comments delivered in his 1933 rectoral address at Freiburg.

Hans-Georg Gadamer states that it is easier to dismiss Heidegger's philosophy on account of his politics; it is harder to continue to engage with it in awareness of Heidegger's culpability. Gadamer writes: "Whoever thinks we can here and now dispense with Heidegger has not begun to fathom how difficult it was and remains for anyone *not* to dispense with him."[14] In my suggestion that there are productive elements in Heidegger's work that Habermas fails to consider, I am firmly following in the spirit of Gadamer's comment.[15]

Habermas begins his lecture on Heidegger's antimodernism with the overarching conclusion that Heidegger's critique of metaphysics, and his assault on modern subjectivism, fails (internally) to overcome metaphysics, remaining within the problematics of the philosophy of consciousness.[16] One question I want to consider is this: Can one transcend the subject without remaining in what Habermas sees as a problematic state of articulating an "abstract Other" of reason?

Habermas further argues that there is an internal connection between Heidegger's engagement with and evaluation of National Socialism and his philosophical work. Habermas attributes Heidegger's *Kehre* to certain changes Heidegger made in the terminology of *Being and Time* during the period of his Freiburg rectorship (1933)—giving concepts from this work a nationalistic interpretation.[17] Contrary to Habermas's interpretation, however, there are many suggestions for the existence of a genuine continuity between the earlier and later work. As I will suggest, although it is extremely important to retain the sense of Habermas's political reading, it is crucial not to dismiss Heidegger's rhetoric of submission—which is bound up with these terminological changes—in the later writings, as simply *reducible* to the political.[18]

Habermas faults Heidegger for failing to elicit any positive elements from the Enlightenment concept of reason. Habermas himself stands almost alone among his contemporaries—thinkers such as Derrida and Foucault—in his staunch commitment to developing the positive potential—the legacy—of Enlightenment critical reason. He faults Heidegger for what he sees as his total dismissal of the concept of reason. However, this means that Habermas's claim is based on the prior assumption that the possibility of critique de facto requires a condition of normativity, which condition Habermas does not see in Heidegger's writing.[19] Habermas equates Heidegger's critique of subject-centered reason with a jettisoning of the possibility of articulating a critical position altogether. For Habermas, Heidegger's critique of modern subjectivism develops one of the enduring themes of modernity since Hegel: the critique of subject-centered reason. Habermas writes:

> Heidegger sees the totalitarian essence of his epoch characterized by the global techniques for mastering nature, waging war, and racial breeding. In them is expressed the absolutized purposive-rationality of the "calculation of all acting and planning."[20]

Habermas calls attention to the fact that Heidegger associates this idea of calculative thinking specifically with the *modern* understanding of Being. This modern understanding sees its beginnings with Descartes and receives its last articulation in Nietzsche. Heidegger says:

> That period we call modern . . . is defined by the fact that man becomes the center and measure of all beings. Man is the *subjectum*, that which lies at the bottom of all beings, that is, in modern terms, at the bottom of all objectification and representation.[21]

According to Habermas, Heidegger's critique of subject-centered reason is problematic. It fails to apprehend any positive potential in self-consciousness and is able to perceive only its authoritarian aspect, while ignoring the existence of a reconciling dimension.[22] In other words, Habermas sees in Heidegger's concept of the modern subject only a "self-perpetuating negativity of ever-increasing activity." Thus Habermas says that, for Heidegger, "the same understanding of Being that spurs modernity to the unlimited expansion of its manipulative power over objectified processes of nature and society also forces this emancipated subjectivity into bonds that serve to secure its imperative activity."[23]

I wish to note briefly one of Habermas's points concerning the consequences of Heidegger's "failed attempt" to transcend the philosophy of consciousness: Habermas takes Heidegger's characterization of "Being" in the later work simply as an hypostatization. He writes:

> Heidegger separates Being, which had always been understood as the Being of beings, from the beings. For Being can only function as a carrier of the Dionysian happening if—as the historical horizon within

> which beings first come to appearance—it becomes *autonomous* to a certain extent. Only Being, as distinguished from beings by way of hypostatization, can take over the role of Dionysus.[24]

Habermas makes the charge of hypostatization—that Being becomes an autonomous element—and uses it as evidence that Heidegger merely supplants a starting point in intentional consciousness—as transcendental, world-constituting consciousness—with a mystical discourse of the event of the withdrawal of Being and its abandonment.[25] In a tone approaching caricature, Habermas cites Heidegger in what he takes to be an example of this abstruse event of Being's withdrawal: that "the staying away of Being is Being itself as this very default." I think it will become clear in my defense of Heidegger, below, what this movement away is supposed to mean.

Referring to Heidegger's relation to the Husserlian development of phenomenology, Habermas claims that in *both* Heidegger and Husserl, "the phenomenological gaze is directed upon the world as the *correlate* of the knowing subject" and that "Heidegger does not free himself from the traditional granting of a distinctive status to theoretical activity, to the constative use of language, and to the validity claim of propositional truth."[26] In other words, Habermas fails to locate any difference between the Husserlian and Heideggerian phenomenological approaches.[27]

Second, and related to this first claim, Habermas criticizes the issue of the indeterminacy of the fate that accompanies Heidegger's account of the event of Being's withdrawal—that the destinings of Being remain undiscoverable. Habermas makes a harsh judgment about the diction and rhetoric of the later Heidegger when he says:

> Because Being withdraws itself from the assertive grasp of descriptive statements; because it can only be encircled in indirect discourse and "rendered silent", the destinings of Being remain undiscoverable. The propositionally contentless speech about Being has, nevertheless, the illocutionary sense of demanding resignation to fate. Its practical-political side consists in the perlocutionary effect of a diffuse readiness to obey in relation to an auratic but indeterminate authority. The rhetoric of the later Heidegger compensates for the propositional content that the text itself refuses: It attunes and trains its addressees in their dealings with pseudosacral powers.[28]

Habermas gives some examples of Heidegger's rhetoric of submission from the "Letter on Humanism," where Heidegger says things like "Man is the shepherd of Being [*der Hirt des Seins*]"; thinking is a meditative "letting oneself be claimed [*Sichinanspruchnehmenlassen*]"; Thinking "belongs to [*gehört*]" Being; it "heeds [*achtet*] the destining of Being"; "the humble shepherd is called by Being itself to preservation [*Wahrnis*: safekeeping] of the truth."[29] Habermas concludes from these examples that, like the language of *Being and Time*, with its "decisionism of empty resoluteness," the later philosophy "suggests the submissiveness of an equally empty readiness for subjugation" and "a blind submission to something superior."[30] Habermas's evaluation is striking in its wholesale devaluation of the later Heideggerian project. He fails to contextualize many of these passages from the "Letter on Humanism," passages in which Heidegger critically examines various conceptions of the human subject—such as Marxism and existentialism—in order to show how, as "humanisms," each fails to locate adequately what is most special about human being—the humanity of man.[31]

A (Qualified) Defense of Heidegger

This constitutes the gist of Habermas's characterization of Heidegger. I would now like to turn to an examination of Heidegger that suggests an alternative to Habermas's valuable but ultimately reductive reading. I want to reexamine the following three claims:

1. that Heidegger's critique of representational and calculative thinking is devoid of critical potential because it fails to locate a reconciling dimension in self-consciousness;
2. that Heidegger fails to overcome the philosophy of consciousness because (a) he remains—with Husserl—tied to a concept of the world as the correlate of the knowing subject and (b) he hypostatizes Being in his later work, conceiving it as an autonomous element, simply the abstract "Other" of reason; and
3. that the passive rhetoric of the later Heidegger has the conservative practical-political consequence of a resigned capitulation to existing authority.

I want to address these issues by arguing their opposites:

1. that there *is* a critical and reconciling dimension to Heidegger's account of technology as calculative thinking—that is, that involves self-consciousness without, however, being reduced to it—something he calls "meditative thinking";
2. that Heidegger *does* move beyond the philosophy of consciousness because (a) he suggests a relation of human beings to the world that moves beyond a conception of the world as the mere *correlate* of the knowing subject and (b) his later discussion of the concept of Being is not a mere hypostatization, where Being exists as an autonomous "Other"; and
3. that the so-called rhetoric of submission is used by Heidegger precisely

in bringing about the first two points, without necessarily conservative consequences.

Characterization of Heidegger's Project

Because many of Habermas's criticisms center on a devaluation of Heidegger's later work, I think it is important to characterize briefly some of the differences between Heidegger's early and later writing. One of Habermas's criticisms concerns the absence of a normative concept of authenticity after *Being and Time*. In addition, Habermas interprets Heidegger's later work strictly in terms of biographical-historical necessity. There are, however, compelling reasons to suggest that there is a continuous thread running through the early to later work: this is Heidegger's commitment to pursuing the issue of the forgotten question of Being.

In *Being and Time*, Heidegger develops an account of human experience that discloses human being as that being for whom its own being is an issue. Heidegger proceeds to emphasize that human beings are temporal beings. He charts an account that shifts its emphasis from human being to Being. That is to say, to the extent that the individual human subject defines itself as aware that its own being is an issue for it, the individual arrives at an (authentic) recognition of its existence as *temporal*. As authentic, human being takes a place in and comprehends Being. Within this basic framework, in *Being and Time* Heidegger analyzes certain pervasive transcendental structures of experience, such as temporality, resolve, being-toward-death, and being-in-the-world. Looking at the early work of Heidegger in relation to the forgotten question of Being, one must ask whether this early focus on *human being* (*Dasein*) allows Heidegger to approach adequately an account of the nature of Being. The pervasive, and transcendental, structures in *Being and Time* are unchanging and final. They provide a grasp of human being-as-temporal, but they are limited insofar as they are structures of *human* experience. They constitute horizons of *human* awareness. The formulation of these horizons—which are the conditions of human awareness—reveals the temporal character of human existence. It does this, however, from the standpoint of human being, and thus still in subjective-transcendental terms.

Heidegger later attempts to comprehend Being more adequately without the transcendental focus of *Being and Time*; actual discussion of human being, the term *Dasein*, figures far less prominently. Heidegger is still seeking to grasp the specifically temporal character of human existence, but to do this in terms of articulating a relation to its ground,[32] rather than simply in terms of the transcendental horizons of human experience.

In this sense, the later work can be seen as an attempt to think beyond the position of subjectivity. But what can it mean to think beyond subjectivity in this way? It is a strange and difficult idea, but one that is suggestively illuminated by a parallel attempt in the work of the poet Rainer Maria Rilke. In 1915, Rilke wrote of the Spanish landscape:

> Everywhere appearance and vision came, as it were, together in the object, in every one of them a whole inner world was exhibited, *as though an angel, in whom space was included, were blind and looking into himself. This world, regarded no longer from the human point of view, but as it is within the angel, is perhaps my real task*, one, at any rate, in which all my previous attempts would converge.[33]

One can think of Heidegger's later work as an attempt to draw close to the character of this world within the angel.

One obvious difference in Heidegger's later writings is that he drops the description "philosophy" and simply characterizes what he is doing as "thinking." He compares this type of thinking to the mode of following a path in a forest (*Feldweg*), implying that philosophical investigation need not begin, and follow, in a strictly teleological sense. One may follow a path that sometimes leads nowhere, yet that occasionally yields unforeseen surprises that come upon one around a bend or turn in the path. A second shift occurs with Heidegger's use and conception of language: he moves from the technical vocabulary of *Being and Time* to a mode of poetic directness, particularly developed through the use of original metaphorical evocations.[34] For example, he concludes the "Letter on Humanism" with the phrase "Language is the language of Being, as clouds are the clouds of the sky."[35]

This is an unusual phrase, yet it has a richly evocative power that questions traditional conceptions of the place of human beings in the world. It does so by pointing out that there are many ways in which the real is, and can be, revealed, that is, not just—as in the Western conceptual tradition—through the medium of representational thinking, a thinking that sets something before human being as an object (*Vor-stellendes*). With this metaphor, Heidegger is evoking a contrast with the usual, specifically *human*, spatial perspective of the sky as "above" the earth, and the clouds "between" the earth and sky. He is challenging this dominant perspective, which begins with the human view of the world (*Weltbildnis*) and suggests that we try to conceive of a complementary conception of this *relation* between human beings and world, one thought from a perspective not beginning with the human but invoking the fact of the surrounding *opening* that allows the world to be revealed to human beings.[36] The earth, as a globe suspended in space, is in turn *surrounded* by this space—in this view, no longer seen from the human perspective, where is the "sky"? How, then, do the clouds relate to this "surround"? This can be more fully developed in relation to Heidegger's views on language and Being and the tension between the movement of revealing/concealing, but this suffices to show that his is a very evocative and incisive metaphor.

A reader of Heidegger such as Habermas, however, seems to have little time or inclination for eliciting the potential of such metaphoric language. Yet it is precisely here that Heidegger can be most valuable in developing an alternative conception of the relation of human beings and the natural world. If the task is to try to begin to think of this relation in a manner other than that of subject and object, then Heidegger's incitement to attend to language—to undergo an *experience with* language—presents one of the most provocative and challenging possibilities. By focusing on the metaphoric (or what Heidegger explicitly calls the *poetic*) force of language, Heidegger is pointing to, and taking advantage of, the liberating quality of language, urging us to think with and through the implications of language. This is most valuable in its power to challenge habitual signification and, hence, habitual modes of representation.[37] It is here that Heidegger has uncovered a truly significant space in pushing toward the limits of the way we (i.e., in the modern period) have traditionally conceived our relation to the real, to the natural world.[38]

Habermas at one point faults Heidegger for a traditional, Hegelian call to reprivilege philosophy. Ironically, he himself betrays just such an inclination: he clearly lends more validity to the "rigorous," terminological, and structured argument of *Being and Time*, rather than to the metaphoric, (philosophically) nontraditional language of the later work. He says of *Being and Time*, "This work—which is argumentatively the most rigourous by Heidegger the philosopher—can be understood as a dead end only if one views it in a thought context different from the one that Heidegger retrospectively arranges for himself."[39] With this suggestion of some of the differences between Heidegger's early and later works, I turn to a defense of his project.

It is easy to cite the later Heidegger in order to make his writing sound abstruse and nonsensical. Concepts such as fate and destiny, which Habermas liberally plays upon, are easily misinterpreted, especially because of the etymological senses that Heidegger often draws upon. The concepts of fate and destiny can indeed sound as if they are linked to something predetermined, demanding "blind resignation," as Habermas suggests. But, for Heidegger, "destiny" is internally linked to the concept of "history." Habermas fails to elicit these connections.

In Heidegger's analysis of the essence of technology, he speaks of "destining [*Geschick*]" as "a way of revealing." In the later works particularly, Heidegger uses the term *destiny* (*Geschick*) in place of *history* (*Geschichte*), playing on the idea of the verb *schicken* (to send), with its resonances of the concept of tradition as *Überlieferung*, or a transmitting and handing down of something from the past. Heidegger wants to counter the traditional concept of history as something simply temporally "past" and therefore alien to us in the present.[40] He stresses that all that is in us from history, *is* itself history. The past is not detached from us, nor is it to be conceived as a mere object of historical research. It should rather be thought of as what is effective (*das Wirkende*) in us. With the notions of fate and destiny (*Geschick*)—to which Heidegger's conception of history (*Geschichte*) is related—Heidegger seeks to draw attention to a particular type of thinking that has come to dominate. This is the mode of calculative-instrumental rationality (*das rechnende Denken*) associated with modern technology.

In the analysis of technology, however, Heidegger locates an alternative mode of thinking that is not calculative. He calls this alternative mode "meditative thinking [*das besinnliche Nachdenken*]."[41] In his elaboration of the nature of this reconciling element—meditative thinking—Heidegger moves toward a conception of the relation between human beings and the world that is beyond the standpoint of subjectivity. In describing this relation to Being, Heidegger can be seen to develop a positive, or reconciling, dimension to what he describes in various ways as a position of "dependence-upon." Through passive locutions, he articulates a conception of dependence/passivity that does not suggest a relation of subjugation to authoritarian force, to an external authority. Heidegger writes in "The Question Concerning Technology":

> The essence of modern technology starts man upon the way of that revealing through which the real everywhere, more or less distinctly, becomes standing-reserve. "To start upon a way" means "to send", in our ordinary language. We shall call that sending-that-gathers [*versammelde Schicken*] which first starts man upon a way of revealing, destining [*Geschick*]. It is from out of this destining that the essence of all history [*Geschichte*] is determined. History is neither simply the object of written chronicle, nor simply the fulfilment of human activity. That activity first becomes history as something destined.[42]

As I have noted, Heidegger says that for modern technology nature is seen as "standing-reserve [*Bestand*]," a term that designates the way we view everything in the world as at man's disposal, standing ready to be ordered by human activity. The essence of modern technology, shown in "enframing [*Ge-stell*]," "is the way in which the real reveals itself as standing-reserve [*Bestand*]."[43] "Enframing," Heidegger says, "is an ordaining of destining, as is every way of revealing."[44] We may note here the significance of the concepts of fate and destiny when Heidegger writes:

> Always the unconcealment of that which is goes upon a way of revealing. Always the destining of revealing holds complete sway over man. *But that destining is never a fate that compels*. For man becomes truly free insofar as he belongs to the realm of destining and so becomes one who listens and hears [*Hörender*], and *not one who is simply constrained to obey* [*Höriger*].[45]

I next want to suggest a positive alternative to Habermas's interpretation of Heidegger's description of fate and destiny as imperatives to which are attached an illocutionary sense demanding resignation.

I will now move to an account of Heidegger's discussion of meditative thinking and the question of Being, which he characterizes in relation to subjectivity as a *mutually determining* relation that lies beyond the subjective-transcendental, in which the world is not simply conceived as the *correlate* of subjective-intentional consciousness. Heidegger examines meditative thinking in relation to the question concerning the character of human being, which, paradoxically, is not a question about human being.[46] He says further that meditative thinking is the way in which human beings are involved directly and immediately in Being.

Thinking is a peculiarly human activity, but it is human in at least two senses. The traditional and usual view of thinking conceives it as a kind of human activity leading to an understanding of objects. In this sense, it is a kind of willing, and is seen as something specifically and merely human. This is a type of thinking that Heidegger describes as ''calculative.'' It is further characterized by an approach to and conception of reality in terms of utility, in relation to our securing an advantage from. In contrast, Heidegger describes another type of thinking, in which he says it is possible that this thought refers *beyond* the human, in the sense that the sphere of reference to human affairs is transcended because subjectivity is not the deciding factor in determining the existence (unconcealment) of the real. This is meditative thinking (*ein Weg des Nachdenkens*), which is essentially related to what Heidegger calls ''releasement [*Gelassenheit*].''[47] Meditative thinking, involving a releasement toward things, is not *grounded in* subjective volition, in human willing. It is crucial to understand in what sense this type of thinking is supposed to refer beyond the human—and this in a sense that is at the same time not just metaphysical (i.e., a ''passivity'' that is not simply the opposite of ''activity''). That is to say, it is crucial to understand that Heidegger can be critical of a conception of human activity that holds the subjective will as ground for what is acted upon; yet this does not require that he bypass the sphere, or possibility, of human activity altogether. Heidegger writes:

> There is then in all technical processes a meaning, not invented or made by us, which lays claim to what man does and leaves undone. . . . *The meaning pervading technology hides itself.* But if we explicitly and continuously heed the fact that such hidden meaning touches us everywhere in the world of technology, we stand at once within the realm of that which hides itself from us, and hides itself just in approaching us. That which shows itself and at the same time withdraws is the essential trait of what we call the mystery. I call the comportment which enables us to keep open to the meaning hidden in technology, *openness to the mystery.*[48]

John Anderson characterizes the difference between representational and meditative thinking as follows:

> To begin to comprehend what is involved in this kind of thinking, we may observe, somewhat negatively, that it does not construct a world of objects. By contrast to representative thinking, it is thinking which allows content to emerge within awareness, thinking which is open to content. Now thinking which constructs a world of objects understands these objects; but meditative thinking begins with an awareness of the field within which these objects are; an awareness of the horizon rather than of the objects of ordinary understanding. *Meditative thinking begins with an awareness of this kind, and so it begins with content which is given to it, the field of awareness itself.*[49]

So, meditative thinking is *open* to what is given. In this sense, Heidegger can state that meditative thinking, without being "about man," concerns "human nature." Meditative thinking is a fundamental characteristic and capability of human nature: it has the character of *openness*. Heidegger has a special name for this higher mode of thinking, defined in relation to the openness involved in it: releasement (*Gelassenheit*). Releasement is a defining characteristic of human being insofar as, according to Heidegger, the true nature of human being involves openness and, through openness, direct and immediate reference beyond man to Being.

Heidegger assumes that meditative thinking, with its focus on the nature of human being, provides a means of directly approaching a recollection of the question of Being, meditative thinking, and a cultivation of this *ultimately* nonvolitional (passive in the reconceived Heideggerian sense of passivity) mode of relation to something beyond itself. Human being, Heidegger says, receives its determining characteristic not from itself, but from something he calls "that-which-regions [*die Gegnet*]."[50] This determining characteristic, the "relation" to *die Gegnet* that is an essential aspect of human being, is related to a type of (nonvolitional) waiting.[51]

As with the two different senses of the activity of thinking (calculative-representational and meditative), Heidegger distinguishes a usual, subjective sense of waiting and another sense that refers the human of necessity to something beyond, and yet intrinsically partaking of, the identity of the subjective. Describing the latter, he says, "Authentic releasement is a releasement-to," as opposed to a "releasement-from." It is directly related to the concept of "waiting." Heidegger says that "in waiting we are released *from* our transcendental relation to the horizon."[52] In the guise of a conversation among three figures on a country path—a scientist, a scholar, and a teacher—Heidegger depicts the notion of releasement like this:

Scientist: This being-released-from is the first aspect of releasement; . . .

Scholar: If authentic releasement is to be the proper relation to that-which-regions, and if this relation is determined solely by what it is related to, then authentic releasement must be based upon that-which-regions, *and must have received from it movement toward it.*[53]

Regarding the notion of "waiting," Heidegger invokes two senses: again, one is a human activity, subjective in the sense of *waiting for* (something) (*das Warten auf etwas*). With this type of subjective waiting, what we are waiting for becomes merely the object of our goals, related to subjective expectations (*das Erwarten*): we wait for things that are expected.[54] The second sense of waiting has a ground in something beyond human being, in a recognition of our *belonging* to what we are waiting for: the sense of this type of waiting can be described as *waiting upon*, rather than "waiting for."[55] In one sense, this type of waiting is a waiting *for* nothing. What we wait upon is, if given, a gift. This idea of a gift element describes an attitude toward the natural environment that contrasts sharply with the traditional calculative conception of nature as standing-reserve. Heidegger writes, "In waiting, we leave open what we are waiting for . . . because waiting releases us into openness."[56] When we are released in this way, an additional element comes into play: Heidegger distinguishes subjective-representational thinking from a mode of thinking he links with gratitude, with "thanking": a sense of indebtedness engendered from an awakened sense of *belonging*.[57] We are, Heidegger says, released to that-which-regions (*die Gegnet*) insofar as we originally belong to it: "Waiting-upon something is based on our belonging in that upon which we wait."[58]

I want to characterize further what is involved in the move from representational thinking to meditative thinking as a shift from the awareness of a world of objects to an awareness of the field of awareness. Meditative thinking is a situation of being open to what is beyond the (specifically human-transcendental) horizon of knowing. The possibility of such an opening, however, depends to some degree upon what lies beyond the horizon of human vision—and upon being open to this "beyond." In this, Heidegger depicts a mutual relation between human being and Being.[59] Seen from within our human perspective, our consciousness of the world of objects is a field of awareness that is unbounded; viewed from within, this awareness has no fixed limits, but only a shifting horizon. Meditative thinking partially consists in becoming aware of the horizon as such and, more significantly, of the nondeterminacy of the human horizon. The human horizon is an opening out, *a standing open in the face of something of which it is not the cause or foundation*. But the very possibility of an awareness, in this explicit sense, of the horizon as an openness exists because the horizon is itself set within an openness of which it is only one "side" (*die uns zugekehrte Seite eines uns umgebenden Offenen*).[60] This openness, in which the horizon of consciousness is set, is what Heidegger calls the "region."

Heidegger describes these two different relations, between human being and the horizon and between the horizon and its "surround [*umgebenden Offenen*]" in the following way:

> **Teacher**: What is evident of the horizon, then, is but the side facing us of an openness which surrounds us; an openness which is filled with views of the appearances of what to our re-presenting are objects.
>
> **Scientist**: In consequence, the horizon is still something else besides a horizon . . . this something else is the other side of itself, and so the same as itself. You say that the horizon is the openness which surrounds us. But what is this openness as such, if we disregard that it can also appear as the horizon of our representing?
>
> **Teacher**: It strikes me as something like a *region*.[61]

With the calculative-subjective mode of thinking, Heidegger says, "Horizon and transcendence, thus, are experienced and determined only relative to objects and our representing them."[62] He then asks the determining question, "What lets the horizon be what it is?" He states that the answer concerns something that is *not usually encountered*:

> **Teacher**: We say that we look into the horizon. Therefore the field of vision is something open, but its openness is not due to our looking.
>
> **Scholar**: Likewise, we do not place the appearance of objects, which the view within a field of vision offers us, into this openness. . . .
>
> **Scientist**: . . . rather, that comes out to meet us.[63]

The crucial point here is that Being (*die Gegnet*; that-which-regions) is not an autonomous element that makes the (awareness of the) field of awareness that Heidegger describes something suprasubjective in the simple sense that it is an elemental thing separate and standing apart from human being. It is *not* hypostatized. Heidegger says that "without man, that-which-regions can not be a coming forth of all *nature*, as it is."[64] The condition of the openness of human existence is *grounded* in the openness of the region; further, the former constitutes only a partial identity of the latter. It does not exhaust it. The most important part of this relation is its feature of mutuality. The identity of human being and Being is to be found in a common—that is, mutual—relation that occurs somewhere *between* (*diesem Zwischen*) human being and Being.[65] It is precisely not a relation between two distinct objects or identities, for, Heidegger says, this aspect of the region, "the coming to meet us, is not at all a basic characteristic of [the] region, let alone the basic characteristic."[66] Heidegger says that "the nature of thinking is not determined through thinking and so not through waiting as such, but through the other-than-itself, that is, through that-which-regions which as regioning first brings forth this nature."[67] The openness of the region, then, which is the *ground* of the opening of human being onto the world, must also be grasped in terms of this dynamism of movement.[68]

This openness of the region is not a vacuum; if it were, it would go unnoticed by human beings. It is in this sense that the previous statements about the *mutuality* of the relation between human being and the world are to be understood. That Being (*das uns umgebende Offene*) *needs* man, and exists partially because of, or in relation to, the nature of human being as open, is central to Heidegger's discussion. He is careful to establish that both are elements of each other, in the sense that *one cannot exist without the other*. Rather than characterizing the region through its relation to us, Heidegger says, "we are searching for the nature, in itself, of the openness that surrounds us."[69] Something similar is brought out in the question, "If a tree falls in the forest, and no one is in the forest, does the tree's falling make a sound?": what is the ontological status of the sound? Following Heidegger, we are incited to think more carefully about the relation between the existence of (the sound of) the tree that is falling and the degree to which this is (or *should* be) conceived as dependent upon human existence. We are led to question more deeply the extent to which the real "exists" in relation to both the facticity of our human existence, that we take part in its existence, and that we also apprehend the real, and construct in part on a basis of this apprehension.[70] More urgently, we should recognize the potential of Heidegger's rhetoric of submission as an alternative to the traditional calculative view that the real exists as a manipulable and exploitable resource, standing ready at human disposal. We must cultivate an awareness of the nobility of a mode of acting that is also a thanking, in Heidegger's sense. This mode of meditative thinking "lets things be" in that it thematizes and acts on behalf of a recognition of its indebtedness, its belonging: we are released to the region insofar as we originally belong to it.

I have described Heidegger's account of the relation between human beings and the world in terms of the region (*die Gegnet*) that is a ground for human thinking, that is, meditative thinking. Heidegger's use of such formulations as "fate" and "destiny," as well as the passive constructions of "releasement-to" and "waiting-upon," is to be understood in terms of what Heidegger describes as lying "beyond the distinction between activity and passivity."[71] Meditative thinking does not involve what is usually considered a subjective act of will because, according to Heidegger's description, one does not will oneself to be open. In this sense, meditative thinking involves an anulling of the will. Yet meditative thinking is not passive in the traditional sense either, because human beings are not open because of neglect or indifference. This passive yielding does not require blind submission, or infringe the ability to act reflectively. In fact, one can avoid the position of becoming a defenseless victim of technology only through an active cultivation of meditative thinking. As Heidegger has said, what calls for thinking in the age of planetary technology can be conceived as an activity different from the usual conception of willing. A brief excerpt from the "Conversation on a Country Path about Thinking" illustrates this:

> **Scholar**: To be sure I don't know yet what the word releasement means; but I seem to presage that releasement awakens when our nature is let-in so as to have dealings with that which is not a willing.
>
> **Scientist**: You speak without let-up of a letting-be and give the impression that what is meant is a kind of passivity. All the same, I think I understand that *it is in no way a matter of weakly allowing things to slide and drift along*.
>
> **Scholar**: Perhaps a higher acting is concealed in releasement than is found in all the actions within the world and in the machinations of all mankind . . .
>
> **Teacher**: . . . which higher acting is yet no activity.
>
> **Scientist**: Then releasement lies—if we may use the word lie—*beyond the distinction between activity and passivity*.[72]

But how we act against activity technologically gone wild is a question Heidegger leaves open.

Conclusion

Certain questions remain to be examined in Heidegger's project, one being the how Heidegger proposes a *mediation* between the now-dominant calculative and meditative modes of thinking. However, my discussion here ends with this account of Heidegger's challenge to a conception of our relation to the real in terms of an activity and passivity in which human beings stand at the center, the view underlying the essence of *modern* technology.

The positive application of this is to be seen in Heidegger's having described as a mutual *relation* what calculative-representational thinking sees in terms of subject and object. As I mentioned above, Habermas's judgment is that Heidegger's rhetoric of submission demands subjugation to an auratic authority, with conservative practical-political consequences. In denigrating *in toto* the relations of submission and dependence that pervade Heidegger's rhetoric in the later writings especially, Habermas seems to belie a transcendental-subjectivist prejudice of his own. I say this simply because Habermas fails to mention any positive potential that Heidegger's account of meditative thinking, formulated through a rhetoric of submission, might offer. He assumes that relations of dependence imply a state of passivity, and are therefore inferior.[73] This inferiority is conceived in contrast to his privileged conception of an activity implicitly connected to certain historically specific ideals of the autonomy and independence of human subjectivity. That is to say, one could locate an additional reason for Habermas's hostility to Heidegger's rhetoric of submission in the fact that Habermas is perhaps himself captivated by the modern conception of subjectivity as one that defines subjectivity in terms of its ability to act upon the world (in effect to master it), rather than in any sense to act with, or be acted upon, in a way that would not necessarily subjugate. I believe it attests to the validity of Heidegger's account of the dominance of the modern technological ethos of mastery that his rhetoric of submission would be so readily misinterpreted.

It is crucial to recognize the positive potential within Heidegger's articulation of the dependence relation between human being and Being, human beings and the world, in at least two respects. First, one rebuts the assumptions of positivism and the philosophy of consciousness concerning the issue of our embeddedness in a social-historical life world. Forces such as language and tradition precede us and limit our complete autonomy. That we are in this sense "dependent" can no longer be conceived as a failing. Second, Heidegger's elaboration of the relation of mutuality and dependence between human beings and nature displaces the dominant calculative-representational model that conceives of man as "lord over the earth." Heidegger calls for a recognition and awakening of the "gift" element expressed in the idea of "waiting upon"; the awareness that our field of vision is open but that its openness is not because of our looking. This awareness has strong practical-political implications for a rethinking of global environmental relations. Present ecological problems are so threatening because of the forgetfulness of man and the dominance of a concept of nature as "standing-reserve" rather than a cultivated awareness of the gift element internal to the relation of human beings and the world. An appropriate commentary on the hubristic blindness to what Heidegger calls the question of Being can be found in his essay on the "Logos": "It is proper to every gathering that the gatherers assemble to coordinate their efforts to the sheltering: only when they have gathered together with that end in view do they begin to gather."[74]

In addition, it is obvious that issues of practical-political application demand our committed attention. But this need not be to an exclusion of the possibility of cultivating an awareness of what Heidegger calls "the mystery"—the question of why there is something rather than nothing, and a fostering of what the Greeks called *thaumazein*, or wonder (*das Erstaunen*). Wonder and curiosity are valuable precisely because they contribute to the recognition that, despite the expansion of modern science and technology into the cosmos at the levels of macro and micro, infinite and infinitesimal, there are some things that are *not* subject to calculation.

Finally, a point concerning the issue of whether Heidegger overcomes the position of the philosophy of consciousness. Insofar as the above discussion was prompted initially as a response to Habermas's evaluation of Heidegger, I have charted a defense of Heidegger on grounds that remain to an extent at a formal level, though I have suggested ways in which Heidegger's distinction of two types of passivity—articulated through submissive rhetoric—has practical consequences for directing action. I have shown that concepts such as destining and waiting are elicited in Heidegger's valuable thematization of our relation to the region called Being, which involves thinking beyond a position of subjectivity. It is not insignificant, however, that Heidegger's rhetoric of submission was enlisted in the rectoral address in the service of defending and supporting a genocidal and fascist regime.[75] Habermas attacks Heidegger in this sense at a more concrete level of particular political engagement. It must remain a disturbing question for Heidegger's thought that these two levels—abstract and concrete, philosophical and political—could be so easily collapsed.

Notes

1. Martin Heidegger, "Die Frage nach der Technik," in *Die Technik und die Kehre* (Pfülligen: Verlag Günther Neske, 1962); translated by William Lovitt as "The Question Concerning Technology," in *The Question Concerning Technology and Other Essays* (New York: Harper and Row, 1977), 22/22. (This work is hereafter cited as *QCT*. For this and all other translated works, cites include page numbers first for the English translation and then for the German edition.)

2. Ibid., 12/12.

3. Ibid., 23/23.

4. Ibid., 26-27/26-27; emphasis added.

5. Owing to limitations of space, I am assuming the reader is familiar with Heidegger's account in "The Question Concerning Technology" and "The Age of the World Picture" in *QCT* and his account of modern science, metaphysics, and mathematics in *Die Frage nach dem Ding* (Tübingen: Max Niemeyer Verlag, n.d.); translated by W. B. Barton, Jr., and Vera Deutsch as *What Is a Thing*? (South Bend, Ind.: Regnery/Gateway, 1967).

6. Heidegger, *QCT*, 14/14.

7. Heidegger writes, "But Enframing does not simply endanger man in his relationship to himself and to everything that is. As a destining, it banishes man in to that kind of revealing which is an ordering. *Where this ordering holds sway, it drives out every other possibility of revealing*. Above all, Enframing conceals that revealing which, in the sense of *poiesis*, lets what presences come forth into appearance. . . . Where Enframing holds sway, regulating and securing of the standing-reserve mark all revealing. They no longer even let their own fundamental characteristic appear, namely, this revealing as such." *QCT*, 27/27.

8. Martin Heidegger, "Zur Erörterung der Gelassenheit: Aus einem Feldweggespräch über das Denken," in *Gelassenheit* (Pfülligen: Verlag Günther Neske, 1959); translated by John Anderson and

E. Hans Freund as "Conversation on a Country Path about Thinking," in *Discourse on Thinking: A Translation of Gelassenheit* (New York: Harper and Row, 1966), 58/9. (This work is hereafter cited as *DOT*.) Heidegger notes of the title of the "Conversation," "This discourse was taken from a conversation written down in 1944-45 between a scientist, a scholar, and a teacher," and Heidegger himself appears to be identified with the teacher.

9. In a qualified defense of Heidegger, I examine those aspects of Heidegger's project that I believe are most valuable for helping to rethink our relation to the *natural* world. While defending Heidegger in this way, however, I am also concerned with Heidegger's inability to accommodate our relations in the *social* world. This issue is one of several that I cannot pursue within this context but that I examine in my forthcoming book, *Hermeneutical Philosophy and the Critique of Modernity*. I concentrate in part on the ontology of language in both Gadamer's and Heidegger's writing, and it is here that I pursue questions concerning intersubjectivity and language, developing a sustained critique of Heidegger. I believe that Heidegger's project offers some profound and valuable insights, however, that are lost to Habermas in the nature of the criticisms he makes. Yet I believe that one must also confront certain aspects of Heidegger's project, and keep in mind the question of Heidegger's affiliation with and support of the National Socialist movement. Habermas is clearly concerned with this issue. As my defense of Heidegger will demonstrate, I do not believe one can gain much insight from Habermas's criticism of Heidegger's rhetoric of submission. One must look elsewhere.

10. Jürgen Habermas, *Der philosophische Diskurs der Moderne: Zwölf Vorlesungen* (Frankfurt am Main: Suhrkamp Verlag, 1985); translated by Frederick Lawrence as *The Philosophical Discourse of Modernity: Twelve Lectures* (Oxford: Polity, 1987) (hereafter cited as *PDM*).

11. It is less that I believe Habermas's readings of Heidegger to be especially nuanced than that I believe a polemical reading such as Habermas's should not go unchallenged for what it too easily dismisses. That is to say, Habermas forecloses too quickly upon certain aspects of Heidegger's attempt to move (us) toward a thinking of the question of Being, or "releasement," through a substantive theoretical articulation of a nonmetaphysical passivity (one not simply opposed to activity, and therefore not ultimately precluding activity, or requiring subjugation to an external authority).

12. That Heidegger's account also has implications for feminist debates concerning the nature of rationality and autonomy is a point I develop in *Earth's Destiny: Rethinking Heidegger* (London: Routledge, forthcoming).

13. Habermas, *PDM*, 155/184.

14. Hans-Georg Gadamer, "Back to Syracuse," *Critical Inquiry* 15 (Winter 1989), 430.

15. For a discussion of Heidegger's involvement, see "Martin Heidegger and Politics: A Dossier," *New German Critique* 45 (Fall 1988): 91-161. For a discussion of the *concept* of the political in Heidegger's writings, see Reiner Schürmann, "Principles Precarious: On the Origin of the Political in Heidegger," in *Heidegger, the Man and the Thinker*, ed. Thomas Sheehan (Chicago: Precedent, 1981), 245-56. For a more broadly conceived examination, which contends that the relation between "thinking" and "acting" is subverted in Heidegger's work, see Reiner Schürmann, *Heidegger, on Being and Acting: From Principles to Anarchy*, trans. Christine-Marie Gros (Bloomington: Indiana University Press, 1987). I wish to acknowledge that one can productively analyze Heidegger's later writings in light of this question of the relation of the political and the philosophical. For an account of this aspect of Heidegger's thought, see Michael Zimmerman, *Heidegger's Confrontation with Modernity: Technology, Politics, Art* (Bloomington: Indiana University Press, 1990); and Pierre Bourdieu, *The Political Ontology of Martin Heidegger*, trans. Peter Collier (Stanford, Calif.: Stanford University Press, 1991).

16. I do not intend to make a judgment here about the validity of Habermas's claim to have moved beyond subject-centered reason—or the philosophy of consciousness. See Jürgen Habermas, "An Alternative Way out of the Philosophy of the Subject: Communicative versus Subject-Centered Reason," in *PDM*, 294-326/344-79.

17. Habermas claims that Heidegger, having made these internal terminological changes, could not—in the later writings—return to the terminology of *Being and Time*. (For example: "*Dasein*" became "*Dasein* of the nation," "freedom" became "the will of the Führer," and the question of Being became the "national socialist revolution as a countermovement to nihilism." Where National Socialism was first seen as a vehicle for the uniting of science and the German university in the "destining" of the German *Volk*, it was later characterized by Heidegger, along with "Americanism," as a symptom of the calculative thinking of planetary technology.) This return, Habermas says, would be tantamount to admitting a mistake in political judgment. This, of course, is the point on which Heidegger has remained disturbingly silent. For this reason, Habermas believes that Heidegger felt *forced* to keep his changed rhetoric.

18. That is to say, the terminology Habermas cites functions, in his reading and criticism of Heidegger, as a criticism of Heidegger's account of passivity and his rhetoric of submission insofar as Habermas claims these lead to a blind and submissive yielding to an auratic authority (see below). Although still problematic, Habermas's evaluation stands on much more solid ground with respect to the specific terminological changes he cites (see note 17). Nevertheless, he should not conflate these specific terminological changes and the conceptualization of passivity/submission in the later writings.

19. The question of normativity in Heidegger's work *as a whole*, but also in the later writings, has not been adequately addressed and warrants further analysis. This question could be pursued fruitfully through an examination of Heidegger's extremely difficult and important concept of ground (*Grund*). See note 31, below.

20. Habermas, *PDM*, 133/159-60.

21. Martin Heidegger, *Nietzsche*, vol. 2 (Pfüllingen: Verlag Günther Neske, 1961), 61; cited in Habermas, *PDM*, 133/160. Habermas chooses this particular description of the modern from Heidegger's *Nietzsche* volumes. Heidegger's essay "The Age of the World Picture" takes this theme as its central focus. Though I cannot pursue this question of intersubjectivity here, note the devaluation of the social realm when Heidegger writes: "Man has become *subjectum*. Therefore he can determine and realize the essence of subjectivity, always in keeping with the way in which he himself conceives and wills himself. Man as a rational being of the age of the Enlightenment is no less subject than is man who grasps himself as a nation, wills himself as a people, fosters himself as a race, and, finally, empowers himself as lord of the earth. Still, in all these fundamental positions of subjectivity, a different kind of I-ness and egoism is also possible; for man constantly remains determined as I and thou, we and you. *Subjective egoism, for which mostly without its knowing it the I is determined beforehand as subject, can be cancelled out through the insertion of the I into the we. Through this, subjectivity only gains in power*." *QCT*, 152. "The Age of the World Picture" was a lecture given on June 9, 1938, under the title, "The Establishing by Metaphysics of the Modern World Picture" in Freiburg im Breisgau. Given the time in which this lecture was initially delivered, Heidegger's inclusion of the ideas of "nation," "race," and so on seems directly aimed at the rhetoric and ideology of the Nazis. It clearly poses a paradoxical situation when compared with the text of the rectoral address, delivered four years earlier.

22. Habermas, *PDM*, 134/161. One of Habermas's claims worth investigating further (and that I pursue elsewhere in an examination of Gadamer's "urbanization" of Heidegger) is whether one consequence of Heidegger's critique of subject-centered reason is that he is no longer able to distinguish between the universal elements—of, for example, Humanism or Enlightenment—and the particularistic "self-assertive representations" of racism or nationalism. Habermas concludes: "No matter whether modern ideas make their entry in the name of reason or of the destruction of reason, the prism of the modern understanding of Being refracts *all* normative orientations into the power claims of a subjectivity crazed with self-aggrandizement."

23. Ibid., 133/160. Habermas's reference here to "emancipated" subjectivity turns on a special sense of this term in Heidegger's critique. Heidegger discusses this type of emancipation or "free-

dom'' as a specifically modern manifestation of ontotheology in ''The Word of Nietzsche,'' in *QCT*. The human subject is, ironically, emancipated in the sense that it is freed *to itself* as the final ground of legislation of truth.

24. Habermas, *PDM*, 135/162; emphasis added. Note Peter Sloterdijk's alternative conception of Dionysus in ''Dionysus Meets Diogenes; or, The Adventures of the Embodied Intellect,'' in *Thinker on Stage: Nietzsche's Materialism*, trans. Jamie Owen Daniel (Minneapolis: University of Minnesota Press, 1989), 50-73.

25. Habermas, *PDM*, 135/162-63.

26. Ibid., 138/165-66.

27. This is a strange conclusion on Habermas's part: ''Because Heidegger does not gainsay the hierarchical orderings of a philosophy bent on self-grounding, he can only counter foundationalism by excavating a still more deeply laid—and henceforth unstable—ground. The idea of the destining of Being remains chained to its abstractly negated antithesis in this respect.'' *PDM*, 138-39/166. There is a wealth of literature devoted to the differences of Husserl's and Heidegger's development of phenomenology. For an original discussion (which also offers a critique of Heidegger's denigration of curiosity in *Being and Time*), see Klaus Held, ''The Finitude of the World: Phenomenology in Transition from Husserl to Heidegger,'' trans. Anthony Steinbock, paper presented at the Goethe Institute conference, ''Heidegger: Philosophy, Art, and Politics,'' sponsored by the British Society for Phenomenology, January 1990.

28. Habermas, *PDM*, 140/168. Also, note the following from *Nachmetaphysisches Denken*, where Habermas writes, ''Allerdings stellt sich dann die Frage, ob die Welterschliessung, das Seinlassen des Seienden, überhaupt noch als eine *Aktivität* begriffen und einem leistenden Subjekt zugeschrieben werden kann.'' Jürgen Habermas, *Nachmetaphysisches Denken: Philosophische Aufsätze* (Frankfurt am Main: Suhrkamp Verlag, 1988), 49.

29. Heidegger, ''Letter on Humanism,'' in *Basic Writings*, ed. David Farrell Krell (New York: Harper and Row, 1977), 189-242. Note that Heidegger's ''Letter on Humanism'' discusses the question of how Being *claims* man: ''But in order that we today may attain to the dimension of the truth of Being in order to ponder it, we should first of all make clear how Being concerns man and how it claims him'' (p. 209) and ''Man is not the lord of beings. He is the shepherd of Being. Man loses nothing in this 'less'; rather, he gains in that he attains the truth of Being. He gains the essential poverty of the shepherd, whose dignity consists in being called by Being itself into the preservation of Being's truth'' (p. 221).

30. Habermas, *PDM*, 141/168.

31. That is, Heidegger's ''Letter on Humanism,'' as a response to Sartre's ''Existentialism Is a Humanism,'' seeks to address this issue of what constitutes the humanity of man. Heidegger criticizes Sartre (as well as Marx) for conceiving human being with an essence, where this human essence serves as a ground. As is well known, Heidegger finds this search for an essence problematic for the same reason that he finds Descartes's view of the human subject problematic (see note 22, above). See also the discussion on this topic by Luc Ferry and Alain Renaut in *Heidegger and Modernity*, trans. Franklin Philip (Chicago: University of Chicago Press, 1990).

32. Heidegger refers to ''ground'' in the sense of ''surround,'' as well as ''foundation'' (*Grund, Boden*) in, for example, *Gelassenheit* and *Der Satz vom Grund*. The concept of ground in Heidegger is extremely complex, and an analysis of the scope and function of this concept exceeds this quite specific reading of Heidegger in response to Habermas's criticisms. However, I do venture here the schematic interpretation that, *contra* Habermas (and many others on this point), there is a principle of reason, articulated through the concept of ground (*Grund*) in Heidegger's writing. This claim is broadly in keeping with my bringing into relief Heidegger's account of meditative thinking as precisely the reconciling dimension that Habermas claims is absent in Heidegger's writing. Justification of such a claim will rest on making further connections with supporting textual evidence that Heidegger's discussion of reason in its metaphysical sense of *Vernunft* must be connected to his ac-

count of ground (*Grund*). When this is examined in tandem with the question of normativity in Heidegger's writing, Heidegger does not appear to be antifoundationalist at all.

33. From Rilke's *Briefe aus den Jahren 1914-1921,* 80; emphasis added. Cited in Rainer Maria Rilke, *Duino Elegies*, trans. J. B. Leishman and Stephen Spender (New York: Norton, 1963), 10. For Heidegger's views on Rilke, see "What Are Poets For?" in *Poetry, Language, Thought*, trans. Albert Hofstadter (New York: Harper and Row, 1971), 89-142. Heidegger believes that Rilke is caught in the problems of substance-metaphysical thinking; Rilke's poetry is restrained by a metaphysics of the will as Nietzsche was restrained as a philosopher of the will. The issue at stake here concerns Rilke and Heidegger's respective views of "the Open." While acknowledging Heidegger's criticism, there are also reasons to look for more in Rilke's poetry than Heidegger's evaluation allows. I cannot pursue this point within the scope of this essay. If Heidegger's account of our relation to the "region" (of Being) is indeed prefigured in Rilke (such as the quotation from this letter suggests), then this would provide one rebuttal to Habermas's charge that what Heidegger is describing (the "open region" of Being) using submissive rhetoric is linked with Heidegger's association with Nazism at a particular historical point. Rilke articulated this view of the world, "no longer from a human point of view," at least a decade before Heidegger. In order to begin to examine this point further, one would need to take into account the function in Rilke's work of the theme of "thwarted love." Heidegger does not consider this central theme, and it functions as a negation of the idea of active fulfillment (i.e., of will) upon which Heidegger's evaluation rests. For a reading of Rilke that goes against Heidegger's interpretation in "What Are Poets For?" see Edith Wyschogrod, *Spirit in Ashes: Hegel, Heidegger, and Man-Made Mass Death* (New Haven, Conn.: Yale University Press, 1985), 189-97. Note also Hans-Georg Gadamer's discussion of Rilke, "Rainer Maria Rilkes Deutung Des Daseins," in *Kleine Schriften*, vol. 2 (Tübingen: J. C. B. Mohr [Paul Siebeck], 1976), 178-87.

34. Of course, language (*Sprache*) assumes a fundamental role in Heidegger's later writings, one formerly reserved for Being. I examine this shift in *Hermeneutical Philosophy and the Critique of Modernity*; see note 9, above.

35. Heidegger, "Letter on Humanism," 242. "Heidegger's Brief *über den* 'Humanismus' " appears also in *Wegmarken* (Frankfurt am Main: Vittorio Klostermann, 1967), 145-94. [*Die Sprache ist so die Sprache des Seins, wie die Wolken die Wolken des Himmels sind* (p. 194).] (The "Brief über den 'Humanismus' " appeared initially as a separate piece in 1949.)

36. I use the word *relation* here, with an emphasis on the feature of mutuality; the idea of two distinct elements is foreign to Heidegger's sense here, and to my use of the term *relation*.

37. I will not develop this point in depth here. I wish to note, however, that one could broaden the implications of Heidegger's emphasis on *experiencing* language by making connections with certain attempts to rethink contemporary ecological problems. These problems, it goes without saying, are of increasing concern. See, for example, the efforts to reconceptualize the moral and ethical dimensions of the distinction of things and nature in itself, *an sich*, from nature for me, *an mich*, that has been proffered by the proponents of "deep ecology." See Arne Naess, *Ecology, Community, and Lifestyle: Outline of an Ecosophy*, trans. David Rothenberg (Cambridge: Cambridge University Press, 1989), especially 47-49, on the difference between a *relational* as distinct from a *subjective* relationship between human beings and the world. Also note Paul Ricoeur's discussion of metaphor and its role in creatively constituting reality in "Metaphor and the Central Problem of Hermeneutics," in *Hermeneutics and the Human Sciences: Essays on Language, Action and Interpretation*, ed. and trans. John B. Thompson (Cambridge: Cambridge University Press, 1981), 165-81. Especially relevant is the discussion of metaphor and semantic innovation.

38. Heidegger's belief that an overcoming (*Verwindung*) of metaphysical thinking is perhaps possible through *having an experience wih language* constitutes an important gain. I believe this gain to be most productive in its capacity to broaden and stretch our conceptualization of the relation of human beings to the *natural* world. The shortcomings of Heidegger's approach in terms of its ability to accommodate the equally important social/intersubjective realm are beyond the scope of this essay.

39. Habermas, *PDM*, 141/169.

40. See Heidegger's account in *Being and Time*, trans. John Macquarrie and Edward Robinson (Oxford: Basil Blackwell, 1962), 424-44.

41. Heidegger, *DOT*, 46/13.

42. Heidegger, QCT, 24/24.

43. Ibid., 23/23.

44. Ibid., 24-25/24.

45. Ibid., 25/24; emphasis added. Note also Heidegger's discussion of human powerlessness in the face of technology in *Gelassenheit*, S. 21-12. Far from advocating that we cannot act, he says, "So wäre denn der Mensch des Atomzeitalters der unaufhaltsamen Übermacht der Technik wehrlos und ratlos ausgeliefert. . . . Aber wir *können* auch Anderes"; S. 21; emphasis added.

46. Heidegger, *DOT*, 58.

47. Ibid., 54/21-22. Heidegger writes, "I would call this comportment toward technology which expresses 'yes' and at the same time 'no,' by an old word, *releasement toward things* [*die Gelassenheit zu den Dingen*]."

48. Ibid., 55/23-24.

49. Ibid., 24; emphasis added.

50. Ibid., 72-73/48-49. "*Die Gegnet*," from its affinity to *die Gegend*, is translated as "region." John M. Anderson has translated Heidegger's noun *die Gegnet* as "that-which-regions," drawing on Heidegger's use of the verb *gegnen* (a sense of movement present in the preposition *gegen*, which can mean "toward" (or "against") and, in Heidegger's own formulation, "*das Gegnende*," *Gelassenheit*, 39. Although this is somewhat ungainly as a term, Anderson's emphasis on the revealing/concealing, to-and-fro character of the mutual relation of human being and *die Gegnet* is important to elicit. Because of Heidegger's emphasis that human being *must* be open to that-which-regions (*die Gegnet*), the *mutuality* of the relation derives from the fact that this openness is not due to our (subjective) willing. See below.

51. Heidegger writes: "Teacher: . . . What is this waiting? / Scientist: Insofar as waiting relates to openness and openness is that-which-regions [*das Offene die Gegnet ist*], we can say that waiting is a relation to that-which-regions." *DOT*, 72/48.

52. Ibid., 73/49; emphasis added.

53. Ibid.; emphasis added.

54. Ibid., 68/42.

55. Ibid., 74/50. ["In der Tat gründet das Warten auf etwas, gesetzt dass es ein wesentliches, und d.h. ein alles entscheidendes Warten ist, darin, *dass wir in das gehören*, worauf wir warten"] (emphasis added). In his 1966 translation of *Gelassenheit*, John M. Anderson adds the prepositions *waiting-for/waiting-upon* to communicate Heidegger's distinction of a merely representational-subjective waiting from one involving releasement. Heidegger himself characterized both as "*das Warten auf etwas*" (e.g., compare *DOT*, 74, and *Gelassenheit*, 50, although he describes the latter as "das Warten auf etwas, gesetzt dass es ein *wesentliches*, und d.h. ein *alles entscheidendes* Warten ist" (emphasis added.)

56. Heidegger, *DOT*, 68/42.

57. Describing releasement as noble, Heidegger writes: "Scholar: Noblemindedness would be the nature of thinking and thereby of thanking. / Teacher: Of that thanking which does not have to thank *for something*, but only thanks for being allowed to thank." Ibid., 85/65; emphasis added.

58. Ibid., 73-74/49-50. [Die Gelassenheit kommt aus der Gegnet, weil sie darin besteht, dass der Mensch der Gegnet gelassen bleibt und zwar durch diese selbst. Er ist ihr in seinem Wesen gelassen *insofern er der Gegnet ursprünglich gehört*. . . . In der Tat gründet das Warten auf etwas, gesetzt dass es ein wesentliches, und d.h. ein alles entscheidenes Warten ist, *darin, dass wir in das gehören, worauf wir warten*]; emphasis added.

59. Its analogue in this instance is *die Gegnet*.

60. Heidegger, *DOT*, 64/37.

61. Ibid., 64/37-38.

62. Ibid., 64/37.

63. Ibid.; emphasis added.

64. Ibid., 83/62.

65. Ibid., 75/51.

66. Ibid., 66/39.

67. Ibid., 74/51.

68. Ibid., 66-67/40-41. Heidegger describes this movement explicitly, which can be translated as "that-which-regions."

69. Ibid., 65-66/39.

70. This question is in keeping with the Kantian separation of the *Ding-an-sich/Ding-an-mich*. We must begin to learn to conceive our relation to nature not just in terms of its referral back to the subjective. As noted above, this question has received attention in recent examinations of the concept of nature in, especially, environmental ethics and so on. Owing to the scope of my discussion here, I cannot pursue these questions further.

71. Heidegger, *DOT*, 61/33.

72. Ibid.; emphasis added.

73. For example, that human subjectivity is "claimed" by Being, or by the forces of language, or tradition, or the like.

74. Martin Heidegger, "Logos" (Heraclitus, Fragment B 50), in *Early Greek Thinking*, trans. David Farrell Krell and Frank A. Capuzzi (New York: Harper and Row, 1975), 62. (I am using David Farrell Krell's translation, which appears on the frontispiece of Heidegger, *Basic Writings*.)

75. See the examples above, but in this specific instance most notably the concepts of "resoluteness" and "resolve" in Martin Heidegger, *Die Selbstebehauptung der deutschen Universität* (Breslau: Gottlieb Korn Verlag, 1934); translated as "The Self-Affirmation of the German University," *Review of Metaphysics* 38 (1985). For further discussion regarding the specific character of Heidegger's Freiburg address, see Karsten Harries, "Heidegger as a Political Thinker," in *Heidegger and Modern Philosophy: Critical Essays*, ed. Michael Murray (New Haven, Conn.: Yale University Press, 1978), 304-28. Also, for transcriptions of various speeches and lectures that Heidegger presented in his year as rector, see "Martin Heidegger and Politics: A Dossier," *New German Critique* 45 (Fall 1988): 97-114.

PART III
Technology and the Arts

Chapter 8

Technical Performance: Postmodernism, Angst, or Agony of Modernism?

Françoise Gaillard

Art and the Loss of Historical Orientation

We are approaching the end of the century. This is not a declaration of millenial inspiration, but a simple fact. Clearly this imminent end is perceived more as just another deadline than as an apocalyptic prophecy or a promise of social and political utopias made flesh. The spirit of the times is in fact neither one of terror nor one of revolutionary messianism, but rather, as we are told in every quarter, one of tranquil hedonism and grasping individualism thirsting for immediate pleasures, all against a backdrop of economic and political liberalism.

Various theories are offered in explanation of this state of affairs. Some analysts, such as Daniel Bell, interpret the phenomenon as an expression of one of the contradictions of capitalism;[1] others see in it the logical evolution of the democratization process ushered in by modernity. But there is general agreement as to symptoms. One of the most obvious ones is without contest the depoliticization that, as Christopher Lasch clearly demonstrates in *The Culture of Narcissism*, combines a disinvestment of the future with a devaluation of the past.[2] This loss of historical orientation results in a generalized indifference evident in the anemic condition of the social body. We are far removed from the radicalism and revolutionary tension of the 1960s. We seem to be suspended in frivolous tranquility, as free from nostalgia as from nihilism; what better term for this state than "cool"?

Art has been sorely stricken by this emotional and ideological disinvestment, and artistic works, which no longer aspire to the criticism of values or engage in reflexive irony, appear to have been whittled down to images whose function and specificity have been undermined. With only the visible surface of art—that is, its often mediocre and rarely surprising images—upon which to base our judgments of it, we have the disturbing impression that nothing is happening in art. The "cooling" of art creates the impression of a crisis of creativity. After a century of avant-gardism that championed the notion of "revolution" in art—linking it, at the height of certain ideological tempests, to notions of political "revolution," as in the *Tel Quel* movement, to name only one—it is not surprising that contemporary art seems to us to be hopelessly bogged down. The market's periodic bouts of fever—due in the main to the fact that art has evolved from commercial value to speculative value—do not obscure the hypothermia manifest in artistic production. It is clear that this leveling out is not unique to the plastic arts alone, but also affects literature and, broadly speaking, the intellectual climate in general; it may be seen in the ubiquitous lamentations on the platitude of the tunes. Indeed, this is one of the most popular themes in cultural commentary these days.

Have the wellsprings of art really run dry? Has the devaluation or depreciation of "revolution," along with its qualitative devalorization, cost art its last resort? Did the decade of the 1980s fulfill the (modernist) prophecy of the death of art? Is it true that art survives only by dint of massive salvage efforts in the media and on the market—a kind of last-ditch cultural CPR? According to some scholars, such as Bell, it is avant-gardism itself, with its imperative to break with the past and seek after novelty, that has exhausted all reserves of invention and condemned art to inexorable reiteration, to the *déjà-vu*. It is true that, for the past decade or so, virtually every forum for the display of contemporary artistic production—biennal festivals, international fairs, exhibitions, and auctions—has been disappointing. But this sense of disappointment is itself suspect. Perhaps it is merely reactive; that is to say, it might be provoked not so much by the painful, and frustrating, death of art as by the death of our criteria for judging art. What we express as disappointment or annoyance is perhaps an inchoate sense of mourning for the negative aesthetics that has characterized modernity. In fact, what is happening today stymies the aesthetic theories of thinkers such as Hegel, Benjamin, and Adorno, thereby forcing us to forsake these frames of reference and to move forward, to revise our aesthetic conceptions. No doubt it is easier to think that we are in the trough of the artistic, literary, or intellectual wave than to come to grips with the lively production in these domains, for this production eludes our philosophical tools. To interpret the lack of originality in formal or aesthetic gestures as a sign of decline is to condemn ourselves to understand nothing of the radical difference that separates a Duchamp from a Warhol, and a Warhol from a Jeff Koons. Simply because art, even with respect to aesthetic forms, is no longer a revolutionary vector, must it be said to be verging on ruin?

This is the difficult question that confronts us today. It is by no means certain that we possess the means to frame the question, much less to resolve it.

The Negativity of Art

No doubt Régis Debray is not wholly unjustified in seeing the pessimistic diagnoses of terminal dullness in which so many contemporary thinkers indulge as a holdover from the prevalent nineteenth-century myth of the condemned artist. This myth gave rise to another myth, which placed a value on the negativity of art and on its role in social criticism. Negativity was espoused at once by the artist as a sort of ethical imperative and by the critic as a principle of legitimation. Negativity thus served a twofold legitimating function. But this negativity could be sustained only by the marginality of the artist and by the socioideological conflicts in which this marginality found its most powerful alibi. Without losing sight of the caricatural or reductive aspect of such a schematic simplification, one might say that there is thus an obvious connection between negativity as a value in modern art and the existence of a sociality based on conflict, or at least one dominated by ideological and social rifts. What can become of art, as it has been defined by modernity, in a society that, against a backdrop of waning social conflict and of the death of ideologies, aspires to democratic consensus? In order to conceive of the role that would devolve upon art in such a context, we would have to be able to conceive of the aesthetic function as shaping a consensus that no longer has anything to do with the consensus upon which Kantian aesthetics is based. (This digression by way of emphasizing the fact that, despite the beliefs of some of our younger philosophers, Kant is of no avail to us in the matter!) The real question is whether, in our shattered societies, consensus can be counted upon to play such an organizational role. Any answer that might be offered now would no doubt be premature. Indeed, there are those who believe that, before tackling the problem of the life of art in a consensual democratic society, we must first rid ourselves of the habit of conceiving of art in the terms established by modernity. Perhaps it is time, they think, to recognize the *ideological*—or, perhaps better, *historical*—character of this conception of art as denial, a conception that numerous theoreticians predicted would lead to the death of art. Perhaps it is time, they suggest, to put an end to the obligatory link between "art" (and, more generally, "culture") on the one hand and "opposition" on the other—a link taken for granted in our approach to art ever since the avant-garde movements. Severing this connection would, according to some, enable us to take a less negative view of what is happening and to become attuned to certain silent processes taking place in art: those gestures tending toward the resacralization or respiritualization of art. Perhaps, some think, the time has come to emphasize the value of art as affirmation and sublimation.

Art and Market Economy

But such thoughts are no more than pious vows. It is difficult to see how art could function as affirmation and sublimation in a society ruled, as is our own, by the market and by individualism. In such a society, one of two situations obtains: either the emphasis on affirmation and sublimation is the doing of the individual artist—in which case the artist is simply one particular case of contemporary individualism, however deceptively this individualism is sublimated by what is left of romantic mythology—or else such an emphasis on affirmation and sub limation is an artistic mutation, and odds are that liberal consensual ideology manages to render futile this will to positivity by subsuming it in the general aimlessness of liberal society. In the view of virtually all our contemporary symptomatologists, our society is characterized by an utter lack of goals, of direction; these lose out to immediate gratification, as Gilles Lipovetsky and others have pointed out.[3] By conforming to the postmodern ethos, with its twofold opposition to *strong affirmation* and to *sublimation*, art is in danger of losing not only its soul, but its essence; it risks becoming the mere purveyor of more or less pleasing images, for the purpose of embellishing our everyday surroundings. No matter which way one turns, it would appear that the future of art, caught between an outmoded modernity and a postmodernity that appears to reject it, is problematic.

But why speak as if this is all in the future? In a certain sense, we are already at this point. We have already entered upon the era of art being reduced to image formation. In saying this, I am not attempting to raise the question of the (aesthetic) value or of the specificity of works that in spite of themselves are affected by this evolution. Rather, I am simply trying to call attention to the current processes by which this value or this specificity is voided. In many cases, artists themselves cannot be held accountable for these processes. Noncynical artists are in fact the first to succumb to this effect of social reappropriation. I am thinking of those, such as Klazen, to name only one, who see their work, born of the radicalism of the 1960s, transformed into images that are received in virtually the same way as we receive mass media images—the only difference being the degree of uniqueness that constitutes the (exchange) value of artworks. In this context it is not difficult to understand the desperate irony that pervades the reactions of an Allighiero or a Boeti, who has chosen to reproduce magazine covers.

Moreover, many artists today are still caught up in a latter-day romanticism, as far as the transcendental value of art is concerned, or in a modernism that is outdated with respect to the place and role of art, that is, its autonomy and its critical import. This gap between mind-sets and realities is perhaps more than anything else cause for concern about the future of art in its present crippled state. It would appear that conceptions of art have never really broken with a pale and flabby modernism, defined as an amalgamation of two musty, atrophied notions: that of autonomy and that of social criticism or even subversion. These outmoded notions notwithstanding, postmodern society has become consensual and "cool." Whether out of some nebulous fidelity to these notions or for lack of any ready ideology to replace them, this modernism, with its values of revolt, is perpetuated, clung to as a superannuated model whose ghost still functions as an alibi. This ghost prevents artists from finding new means of justifying their activity, their role, their stylistic choices—in short, from inventing a culture and a language that could express the ethos of the period while remaining distinct from it. Taking refuge from modernity and its spirit in an ineffectual nostalgia that is most often barely conscious of itself, conceptions of art run the risk of being cut off from the reality of art, that is to say, from an understanding of the processes of banalization, of insignificance, of "kitschization" in which art is enmeshed in this epoch.

Art and Everyday Values

It is toward desubstantialization and loss of meaning that art is being led by postmodern culture, which Lipovetsky with good reason defines as "the end of the divorce between artistic values and everyday values."[4] Lipovetsky has also clearly demonstrated how the absence of contradiction was already leading toward the vacuum that Cioran predicted, as early as the late 1950s, would be the price of freedom. "Liberal society," wrote Cioran, "by eliminating 'mystery,' 'the absolute,' 'order,' and lacking true metaphysics as well as true policy, throws the individual back upon himself, all the while divorcing him from what he really is, from his own depths."[5] The outcome of this prediction is plain in the domain of art. Every individual, secluded in his or her singularity, which is no longer dialectically linked (as it used to be in the romantic and avant-garde eras) to a transcendence, wears him- or herself out in restless pursuit of a personal style, which might well be referred to as the individual's "look": it is no accident that the term invokes the Madison Avenue media world. Can this phenomenon be stopped or even reversed? For this to happen, there would have to be a renewal of critical radicalism—before the collective encephalogram of our society and its culture bottoms out once and for all. At the present, however, there is no indication of a potential for such a renewal in art: because of its economic stakes, art is, of all instances, the one *least* qualified to provide the tonic electroshock. Why, you may well ask, pine after such a renaissance? Isn't such wishful thinking just one more symptom of the old "modernist" reflex that, according to the logic of our time, we should have rid ourselves of?

All symptoms converge to indicate that we have entered upon a new era in the history of democratic societies. Deciding whether this is the result of individualism carried to its logical conclusion, as Lipovetsky suggests, or of capitalism torn apart in the clutches of a profound cultural crisis, as Bell would have it, is less important here than becoming aware of a (slow?) mutation whose effects have suddenly made themselves known—as if we had all at once awakened to a radically different intellectual, cultural, and social landscape that resists interpretation according to our extant frames of reference. In such a situation, how could this indifference to what used to hold meaning for us be conceived in any but a reactive or nostalgic mode—that is, involuntarily, in terms of a loss of substance or disappearance of meaning, in short, in terms of a "void" or "vacuum"? For once, we are at a loss; we have no ready-made answers at our disposal to account for the new reality. We must therefore proceed by collating the convergent symptoms, however heterogeneous they may be, however they may resist our attempts to make sense of them in a coherent diagnosis.

Modern and Postmodern

The term *postmodernity* is itself a skimpy cover-up for our poor comprehension of these phenomena. This epithet translates our consciousness of a whole slew of changes that are particularly visible in the cultural, and most particularly the artistic, domain. Thus everything that seems to run counter to what was yesterday or the day before considered "modern" is declared to be postmodern. Postmodernity seems still to be seeking itself somewhere between a rationalized rejection of modern style (or nonstyle) and a freewheeling abandon, by turns liberal, skeptical, and cynical, giving itself over to the arbitrary rule of "anything goes."

I am inclined to agree with Lipovetsky when he writes:

> The postmodern age is by no means the age in which modernism comes to a head, in a frenzy of libidinal drive; rather, I tend to think that quite the opposite is true: the postmodern period is the cool, blasé phase of modernism, a trend toward the custom-built humanization of society, the development of fluid structures geared to the individual and his desires, the neutralization of class conflicts, the dissipation of the imaginary of revolution, mounting apathy, narcissistic desubstantialization, the cool reinvestment of the past.[6]

Granted, the notion of postmodernism is not always clear, because it lumps together very diverse symptoms, but an instructive reading of the phenomenon is offered to us by architecture, which is moreover where postmodernism saw the light of day. We know that in this field, the term *postmodernism* is used in order to qualify those accomplishments that turn their backs on modern functionalism, and that are characterized by the use of ornamentation, by stylistic eclecticism, and by borrowings from various traditions. It should be added that postmodernism involves only a simulacrum of revitalization, for before the lively forms of earlier times are recycled, they first undergo a process of neutralization, of banalization, and of desymbolization: this process is the postmodernist gesture par excellence. Art and architecture suffer from the same syndrome: social liberalism, or the devitalization of meaning.

The Return of the Sign as Expression

From a perusal of contemporary journals in the fields of architecture or of contemporary art, it becomes evident that the sign as expression—and even more so the symbol—makes a forceful comeback. But it would be inaccurate to see herein the symptom of a tendency toward resymbolization in art. On the contrary, this resurgence of sign and symbol is an indicator that art has been emancipated to such a degree that it can dare to use the symbolic without running the risk of regression. Let us take an example that is seen as an architectural success: that of the Institut du Monde Arabe, on the Quai Saint Bernard in Paris. It is a magnificent building of clear, pure lines, one of whose facades is ornamented with diaphragms that are opened and closed by a photomechanical device. Now we begin to understand. Architecturally speaking, the Middle East is represented by, among other things, the *moucharaby*, a panel that serves to screen women from view while allowing them to observe. This characteristic element is metamorphosed, modernized, and Westernized by the use of glass and steel; reduced in scale to a decorative diaphragm, it now serves only to add rhythm to the southern facade of the building. The *moucharaby*, this highly symbolic architectural element coming from a culture that enjoins its women to "see without being seen," is here transformed into a simple motif of merely ornamental value. This is a typical example of the denaturing or deculturization of signs, effected under the cover of a return to tradition, that is so characteristic of postmodernism. A formal element charged with meaning in its original context is diverted from its original function in order to serve solely as ornamentation. Emptied of meaning, and therefore emptied also of its emotional or phantasmatic punch, it is renegotiated as a purely decorative effect.

Every epoch witnesses periods of icing over. With postmodernism we have entered the period of cold symbolism. And if the imagination of the period profits from this process nonetheless, it is because the period has already invested in technical performance, the precision of forms, the sharpness of execution, the nobility of materials: in short, a perfection of the whole. Corresponding to "cold" symbolism is a purist imagination!

Cold Symbolism

What do I mean by a *cold symbolism*? I refer to what happens when a symbolic object is decontextualized, isolated from both its formal and its cultural frameworks, cut loose from its social, historical, and stylistic mooring. The reappropriation of forms detached from the aesthetic or cultural whole in which they had both function and meaning is hardly a new phenomenon. Didn't cubism draw in part from the treasury of forms offered by primitive arts? How then is postmodernism different? The difference stems from the fact that cubism was seeking to revitalize aesthetic forms by a return to primitive or archaic sources, whereas postmodern architects borrow forms in order to produce aesthetic effects. Their interest in tradition does not come from a desire to replenish or reinvest the forms, but from a need to find "gimmicks," in other words, novelties that confer singularity. The postmodern architect is therefore not looking for ideal or original forms, but for ideas of form or, rather, for forms in the service of an idea. We must not be taken in by appearances. The postmodern reappropriation of formal elements, the recourse to heavily symbolically charged references, is in no way intended as a symbolic revival of the qualities, now absent, that modernism had originally bestowed upon the these forms; postmodern buildings, whose forms no longer express their functions, are, and remain, devoid of these symbolic qualities. The instrumentalization of the symbol, which means its decorative or "quotable" use, is the surest way to cancel out its value, which lies precisely in its symbolic function. With postmodernism, we have clearly entered a new era, one that turns its back upon nostalgia. "Retro" is not backward looking. Yesterday's symbol is today no more than a signal that, in the tightly woven urban web, indicates the purpose of each building. An Institut du Monde Arabe can be recognized by its pictogram of *Arabness*. The times are no longer those of *symbolism*, but those of *signalism*. The art is that of facades, of surfaces, where cultural imprints and the appeal to memory no longer open any window onto a beyond of signs. The same may be said for painting.

Is the process of desymbolization that was begun by critical modernity completed by postmodernity, or is it still in progress? Can art today reabsorb the symbols, images, and forms bequeathed upon it by even the most antitraditional of traditions such as functionalism in architecture, because these no longer symbolize anything, because they are signs without memory, "forgetful of themselves," to borrow Adorno's term. All taboos can be lifted, especially those that weighed upon the heritage of the past. Its treasury, reduced to a pure repertory of formal devices without any trace of a reminder of their origins, can be put back into circulation. All one can do now is go fish from the toy box. Another example: recently, a catalog devoted to Le Corbusier presented itself as a lexicon of terms and of forms, a dictionary of technical contrivances and stylistic discoveries. What has disappeared is the unifying project—that which exercised its power of integration to confer upon the various elements a unity of style. What remains are the scattered pieces whose aesthetic value is now problematic—can this value in fact exist independent of the whole? This aesthetic value is recycled, redirected, ad libitum, toward ornamental ends. It is hardly surprising, then, that so many contemporary achievements give us the impression of being nothing but superficial decoration.

Will the wanderer notice any difference between the stage and the street, if the aim of both is simply effect for its own sake? For example, we no longer construct a monument; rather, we build monumentally with dizzying, dazzling means. We are looking now for what, in the history of styles, used to represent monumentality. The column, being one of the most obvious signs, is recycled, preferably as a portico, because repetition is one of the (easiest) secrets of grandeur. What is striking, despite the heterogeneity of plastic solutions created in order to misdirect the amateur as well as the art historian, is the uniformity of aesthetic postures and practices; this consistency is what confers heteroclitic postmodern works with a certain style. But when style is all, the posture of mannerism is not far off, nor is kitsch. Postmodernism, or the *kitschization* of art, or simply the *kitschization* of culture and social life.

Kitschization

According to some critics, modernity is to blame for all this. Indeed, this kitschization is explained simply as the effect of the banalization of aesthetics, responsibility for which is ascribed to avant-garde policies of aestheticizing the banal. Blame lies with the avant-garde either because, by dissolving aesthetic norms, the avant-garde paved the way for the creation of the oft-propounded "anything goes" thesis or because, by making innovation and surprise the supreme artistic values, the avant-garde movements ended up exhausting the creative powers of negativity. The thesis of the exhaustion or burnout of modernity is quite widespread. Those who subscribe to this thesis call to witness the sterility of the avant-garde movements, which they see as having been given over wholesale to an escalating sensationalism that must sooner or later—but inevitably—wear them out.

The Crisis of Art

However, the crisis of art or, at least, as Paz puts it, of "the idea of modern art," cannot be adequately explained by this recourse to the so-called logic of the avant-garde, seen as separate from its historical context. A better explanation would take into account two social phenomena masked by this avant-garde thesis: (1) the democratic expansion of the artistic domain, resulting in the banalization of avant-garde inventions by their reproduction on a smaller scale and their circulation for the purposes of exploitation by the mass media and the market; and (2) the temporal acceleration that, by abolishing the latent period between the subversive avant-garde proposition and its social reappropriation, has annihilated the subversive power of art. All this is to say that it is false (and above all suspiciously convenient) to think that modernity is dead from exhaustion—if indeed it is dead! And it is even more false to justify postmodernism's pathetic attempts at formal renewal by arguing, as does Bell, that because "experimentation had become tiresome, and transgressions were no longer transgressive," it was necessary to find something else, a second wind, a new source of vitality, and that these were no longer to be found anywhere but in the cultural patrimony scorned by modernity.[7] To think this way is doubly erroneous. The first error, a methodological one, is believing in the autonomy of the history of art (the autonomous logic of avant-gardism); the second, a diagnostic error, is seeing the postmodern attitude as a revitalizing return to the roots of art.

Perhaps there is also a third error, one that lies in the way modernity is seen. Certainly it is a vague, slippery notion, nearly impossible to grasp in a historical perspective, since every period enmeshes it in a different web of significations and functions. But in this confusion, which only increases if approached in terms of historical period, two things seem to characterize what is generally understood by the term *artistic modern*: autonomy and a will to break with tradition. The avant-gardist radicalization of these elements is what gave birth to the late nineteenth-century notion of modern art. But if we step back a little, we can see that the avant-garde adventure is but the final phase in a long process of rationalizing man's relations with the world, a process begun in the Renaissance and culminating in the worldview of the Enlightenment. The doctrine of "art for art's sake," formula for the autonomy of the artistic sphere in the nineteenth century, is therefore inscribed both in the logic of *objectification* of the concept of the Beautiful (which owes its theoretical justification to Kant, although it originated long before him) and, simultaneously, in the logic of *subjectification* of creative inspiration.

For those who subscribe to the exhaustion theory, the consequences are self-evident. By opening itself to subjectivity, art accedes to an imperative for originality, which, taken to its extreme by avant-garde movements, is transformed into an obsession with innovation and revolution. Awaiting (modern) art at the end of this trajectory is death by impoverishment of its inventive resources. This schematic vision, seeing modernity as exclusively caught up in the theatrical gesture of perpetually breaking with tradition, too easily overlooks what modernity owes to tradition. What is forgotten above all is the *ideological* and *political* meaning of the gesture of rupture with the past. By limiting the innovative and revolutionary imperative to the level of aesthetic effects (the invention of forms) and not extending it to the level of a critical political agenda, we may well fail to understand that what killed modern art was the disappearance of the agonistic, antiestablishment energy of contemporary society. The death of modern art is only one sign of the death of all critical functions in consensual liberal society. This is no small matter.

Artistic Creation and *Fin de Siècle* Melancholy

If in this *fin de siècle* period we have cause to worry—but this is a nonchalant kind of worry. In keeping with the tenor of the times, it lies in the latent depression, the smoldering melancholy, the creeping senility that preys upon democracies. The whole sphere of artistic creation has been stricken with this affliction. Of course there is no lack of voices to warn us against the lethal charm of liberal societies, which submerges them in a pernicious lethargy verging on anemia. But what is the use of standing on a soapbox, along with certain nostalgic intellectuals, to strike up the same old "Democracy Blues"? This feeble attempt to rouse the slumbering social body to wakefulness is doomed, like all other such attempts, to fail. It will occupy its allotted slot in the mass media along with other soothing or innocuous spiels, and its proximity to these will neutralize it. And yet, the prognosis can hardly be dismissed.

Will postmodernism turn out to be a symptom of the aging of modernity or, rather, the name given to this geriatric depression? Besides, how can we not see that what we call "democracy" in postindustrial societies is actually a collective apathy that has infiltrated the social body, not only unchallenged, but aided and abetted by a devalorization of the qualitative values of democracy—or rather, not by their devalorization but by their devaluation, their deflation through the removal of any meaningful contents—just as meaning has been drained from everything else? Our epoch is characterized not by reversals of meaning but by the irrelevance of the question of meaning. This, no doubt, is the reason we feel so bewildered. It is as if the place where meaning is formed had become a black hole, a great maw of antimatter that swallows up whole worlds. Now, with no concern for meaning, and least of all for the meaning of meaning, can artistic activities—to speak only of art for the moment—have any meaning?

This depression of values, not to be confused with their depreciation, clearly affects the social field as a whole. Tolerance has turned to indifference, freedom of expression to the neutralization and banalization of messages, the search for consensus to flattening and homogenization. "Cool," the virtue of social liberalism, is a drastic drop in temperature, threatening to plunge the social body into an irreversible coma. Let us make no mistake about it: the peaceful coexistence of ideas and styles that we are now witnessing is less an index of any maturity on the part of the "consumers" of freedom of thought than it is an indication of the voiding of the underlying concepts and principles of this liberal imperative.

This appears all too clearly in the overwhelming responses to a survey of some fifty artists conducted for the newspaper *Le Monde diplomatique* by Yves Hélias and Alain Jouffroy. This survey was recently published with the title "A Portrait of the Fin de Siècle Artist." The unemphatic, muddled reactions to the important question (important especially for a society that considers freedom a crucial value!) as to the role of art in the defense of freedom plainly illustrate that this defense—taken for granted to the point of blurring indistinguishably with the social consensus—is devoid of creative democratic energy. This clearly means that, at present at least, we must not count on artists to provide us with any compelling symbolic anchor for democratic principles in the collective imaginary. And, indeed, why should we expect that artists would be immune to the generalized phenomenon of acquiescent inertia with regard to values?

If certain intellectuals have sounded the alarm, it is because they fear—and the examples I have just cited only argue in their favor—that the disinvestment of political and ideological values will lead to indifference concerning democracy as a value, particularly as the threat of totalitarianism seems to have abated. How is it possible to reawaken the passion for democracy, which is at the heart of the modern project, and which remains the sole bulwark against all atavistic impulses to seek meaning in religion, nationalism, populism, racism, anti-Semitism, or other regressive movements? Whereas art, in its modern definition, would have taken up this challenge, I fear the issue is not even present as a theme of art in its postmodern phase.

For some time now modernity has been in what I would call, somewhat paradoxically, its "disenchanted" phase. This tendency toward neutralization or banalization, toward the indifferentiation of symbolic contents, whose architectural manifestations I have tried to evoke, is ubiquitous. One, perhaps the most immediately perceptible, aspect of this phenomenon, and the most painful for an intellectual to contemplate, is no doubt the dehierarchization of works of thought or of art, all of which have been rendered equivalent in the undifferentiated eye of the consumer. And let it not be said—as Lipovetsky would have it—that this indifferentiation, this lowest-common-denominator treatment of tests and images, represents the fulfillment of the egalitarian democratic ideal!

It is not difficult to understand—although one need not automatically follow suit—why conservative thinkers such as Bell consider postmodernism simply as another name for the moral and aesthetic decadence of our time: postmodernism, or the twilight of the West. However I cannot accept the moralistic and teleological vision implicit in this view. For this reason, I personally see postmodernism as a malignant, and epidemic, form of anemia that has attacked modernity, unless it is simply a moment in the normal evolution of the modernist enterprise—perhaps even its execution, in both senses of the term, given that it brings to an end the negative dialectic that was the mainspring of the artistic and intellectual vocation. If art is a fruitful place to observe this evolution, it is not only because the idea of modernity is closely related to the history of European art, as Habermas has reminded us; it is because art has directly sustained the perverse effects of the modernist process. The (modern) bias toward the aestheticization of the banal has in fact led to a (postmodern) banalization of aesthetics. Hegel prophesied the generalized aestheticization of existence, but how could he foresee that it would take place through the dehierarchization of values, which would claim art as its first victim?

We are living in the bizarre aftermath of modernity, and art is at present caught in the contradictions of the transition from one era to another; but if it manages, while becalmed in the doldrums of consensualism, to keep its sights fixed upon the goal of ironically integrating the various aspects of life, then modernity still has some surprises in store for us.

Translated by Jennifer Gage

Notes

1. Daniel Bell, *The Cultural Contradictions of Capitalism* (New York: Basic Books, 1978).
2. Christopher Lasch, *The Culture of Narcissism* (New York: Norton, 1978).
3. Gilles Lipovetsky, *L'empire de l'éphémère: la mode et son destin dans les sociétés modernes* (Paris: Gallimard, 1987).
4. Ibid., 8-9.
5. Emile M. Cioran, *History and Utopia*, trans. Richard Howard (New York: Seaver, 1987).
6. Lipovetsky, *L'empire de l'éphémère*.
7. Bell, *The Cultural Contradictions of Capitalism*.

Chapter 9

The Technology of Death and Its Limits: The Problem of the Simulation Model

Scott Durham

Mechanization and the Resistance of the Organic

One of the early problems confronted by modern industrial technology was the mechanization of mass slaughter—a process that, as Siegfried Giedion's work suggests, is in many respects emblematic of the experience of death in high modernity.[1] In other branches of industry—textiles and machine tools, for example—the advantage of the machine lay in the replacement of the irregular and discontinuous movement of the human hand with the unvarying and constant rotation of mechanical motion. "In its very way of performing movement," writes Giedion, "the hand is ill-fitted to work with mathematical precision and without pause" (p. 46). The demands of mass production pushed industrial technology to detach the performance from the performer: to free movement from the internal limits imposed by the human gesture and the organic unity that had been its support. In areas where the raw material was already inanimate, this goal was readily achieved, but the slaughter and dismemberment of animals posed special problems, for, as Giedion writes, "whenever mechanization encounters a living substance, . . . it is the organic substance that determines the law" (p. 195).

The first problem was of a straightforwardly mechanical nature: required was a machine sufficiently supple to match the living animal's unpredictable variations of shape and dimension, so as to kill and "disassemble" it without spoiling the meat itself. But innumerable attempts to perfect such a mechanism proved unsuccessful:

> The transition from life to death cannot be mechanized if death is to be

> brought about quickly and without damage to the meat. What mechanical tools were tried out proved useless. They were either too complex or outright harmful. . . . Only the knife guided by the human hand can perform the transition from life to death in the desired manner. For this operation craftsmen are needed who combine the precision and skill of a surgeon with the speed of a piece worker. It is established how far and how deep the throat of a hog should be pierced. A false stroke injures the meat product. And it must be done quickly—500 hogs per hour. (pp. 243-44)

Second, and perhaps more troubling, was the intransigence of the animals themselves. Hogs, for example, are "quite likely to become suspicious" and will "perhaps . . . even resist being driven" into a narrow passageway (p. 233). For this reason, any number of mechanisms and ruses—including the use of a "decoy hog," happily consuming slop within the death house into which it was supposed to lure its fellows—were brought into play, in order to avoid the disruption of production by those victims critically inclined enough to take exception to the process.

Here, the resistance of the organic to mechanization no longer appears merely as a matter of passive nonconformity, which might be resolved by the inventive stroke of the engineer alone: it appears as a subject's refusal to be seduced into collaborating with its exterminators. But in both cases, the organic resists mechanization as something imposed upon it from without. In Giedion's invocation of the organic world, industrial technique thus confronts its external limit—a limit that, although it might be pushed back, still remains inviolable. Indeed, writes Giedion with a certain satisfaction, in the struggle between engineer and hog, "the engineer did not emerge victorious" (p. 232), for to this day the actual killing, boning, and the skinning of the pig still requires a direct confrontation with other beings of flesh and blood. The image Giedion offers of a modern meat-packing plant is a contradictory one; it mingles technical precision with sacrificial agony:

> Killing itself . . . cannot be mechanized. It is upon organization that the burden falls. In one of the great packing plants, an average of two animals is killed every second—a daily quota of some 60,000 head. The death-cries of the animals whose jugular veins have been opened are confused with the rumbling of the great drum, the whirring of gears, and the shrilling sound of steam. Death cries and mechanical noises are almost impossible to disentangle. Neither can the eye quite take in what it sees. On one side of the sticker are the living; on the other side, the slaughtered. Each animal hangs head downwards at the same regular interval, except that, from the creatures to his right, blood is spurting out of the neck-wound in the tempo of the heart-beat. In twenty seconds, on the average, a hog is supposed to have bled to death. It

> happens so quickly, and is so smooth a part of the production process, that emotion is barely stirred. . . . One does not experience, one does not feel; one merely observes. It may be that nerves that we do not control rebel somewhere in the subconscious. Days later, the inhaled odor of blood suddenly rises from the walls of one's stomach, although no trace of it can have clung to the person. (p. 246)

This description is striking in its juxtaposition of each animal's individual struggle for life with a vast apparatus designed to reduce the victim's organic uniqueness to the interchangeability of an industrial product. More striking still, however, is that this juxtaposition can appear only retrospectively to the contemplating subject, who must reconstruct what could not be mastered on the spot. The organic rebels against the neutrality that governs the gaze of the observer, demanding that he or she pay homage to its sacrifice: but that observer him- or herself is incapable of experiencing this last protest of a fading life except through the traces left by its annihilation. The historian's backward gaze, which attempts to redeem the organic at the moment of its liquidation, is at the heart of Giedion's history of industrial technology, over which Benjamin's "angelus novus" might be said to preside.[2] The death of the animal in its irreducible living presence appears only retrospectively—illuminated by its unavoidable collision with technological progress. "The greater the degree of mechanization," writes Giedion, "the further does contact with death become banished from life. Death is merely viewed as an unavoidable accident at the end" (p. 242).

Giedion's insistence on the animal's fading life is thus decidedly "untimely" in a way that links his writing of history to the nostalgic and restorative impulses associated with literary modernism.[3] It depends on the historian's determination to salvage the organic through its afterimage, against the grain of a development that increasingly conjures away not only the organic, but even the experience of death that marks its loss. Such an experience presupposes that something like an experience of the "organic" is still at least imaginable. Today, however, as Giedion himself foresaw, this experience seems to have been largely displaced by one (now christened "postmodern") in which the limit separating the organic from the inorganic, like that between the original object and its serial reproduction, has been increasingly effaced—along with the specifically modernist forms of protest and dreams of redemption to which it once gave rise. The emblem for the contemporary experience of death is, I would suggest, no longer to be found in Giedion's slaughterhouse, but in the simulation models evoked by J. G. Ballard in his novel *Crash*.

The Simulation Model and the Death of the Subject

In Ballard, the "accident" of death itself emerges as a product of triumphant industrial planning and technology. Ballard describes the following scene: at an automotive testing site, the Road Research Laboratory is demonstrating its techniques for simulating accidents. Before a group of spectators, a family of four mannequins is subjected to a high-speed collision. A motorcycle—with a mannequin of its own shackled to the handlebars—plunges into the cabin of the car. Stripped of its domestic carapace, the family within suffers the injuries for which it has been destined. Lacerations, dismemberments, decapitations: all dutifully appear at the points on their heads and torsos that the engineers have marked in advance with coded dyes and cryptic symbols:

> Already, as the vehicle moved back under the impact of the collision, the four occupants of the car were themselves moving toward a second collision. Their smooth faces pressed on into the advancing windshield as if eager to see the chest glider soaring up the bonnet of the car. Both the driver and his woman passenger rolled forward to meet the windshield, touching it with the crowns of their lowered heads at the same moment as the motorcyclist's profile struck the glass. A fountain of spraying crystal erupted around them, through which . . . their figures were taking up ever more eccentric positions.[4]

Ballard's style, in mobilizing the automatisms of pornographic fiction alongside the vocabulary of a technical manual, conveys an apathetic fascination. It thereby mimics the stance of the simulated crash's witnesses themselves, who, calmly turning away from the still-smoking metal, have eyes only for its reproduction on the screen:

> Helen Remington held my arm. She smiled at me, nodding encouragingly as if urging a child across some mental hurdle. "We can have a look at it again on the Ampex. They're showing it in slow motion." (p. 125)

Before this displaced repetition of the simulated accident the crowd of visitors indulges in that ambiguous pleasure familiar to every viewer of CNN: the pleasure in simultaneous contact and absence that Jean Baudrillard has so well described as the "thrill of the real, or of an aesthetic of the hyperreal . . . , a thrill of alienation and of magnification . . . at the same time."[5]

> The audience of thirty or so visitors stared at the screen, waiting for something to happen. As we watched, our own ghostly images stood silently in the background, hands and faces unmoving while this slow-motion collision was re-enacted. The dream-like reversal of roles made us seem less real than the mannequins in the car. I looked down at the wife of a Ministry official standing beside me. Her eyes watched the film with a rapt gaze, as if she were seeing herself and her daughters dismembered in the crash. (p. 128)

The crowd is eerily calm as sentence is carried out on these "copies" that surpass the spectators in reality even as they precede them to their deaths. But on whom has sentence ultimately been passed? It is this question that continues to trouble the crowd. The scenario of their deaths has been narrated in the smallest detail: the severity and location of their probable injuries at a given velocity have been demonstrated, the image of their individual deaths has been recorded, and could be both verified and repeated an infinite number of times. No greater triumph of planning is imaginable than this moment in which the accidents of death itself emerge as products of the operational universe. The one moment that we formerly took to be the most unmasterable, the most unpredictable, but at the same time for each of us peculiarly "mine"—the moment of "my death"—now appears as the product of social engineering, coupled with the most advanced technology. The industrially engineered simulacrum of their deaths, which emerges from this process, appears to the crowd as its truth, while they have themselves been relegated to the status of mere "images" in the background. Thus, the simulacrum appears as the truth of the original in the strongest possible operational sense: as a simulation model—a copy that not only precedes but generates its original.

In his discussion of such simulation models, Jean Baudrillard has emphasized not only the extent to which they efface any notion of an original or auratic object that would precede its reproduction, but also how, as a consequence, they undermine the autonomy and individuality of the consuming or spectating subject that submits to the rules of its operation. Baudrillard starts his account conventionally enough, with a discussion of the demise of the auratic work of art as first described by Walter Benjamin in "The Work of Art in the Age of Mechanical Reproduction."[6] Benjamin showed how, as techniques of mechanical reproduction gradually absorbed the original work itself (particularly in the cases of photography and film, where reproduction emerged for the first time as clearly inseparable from the production of the original), the auratic presence and authority of the original work was rapidly worn away; but Benjamin himself, Baudrillard goes on to argue, could not have anticipated the extraordinary hegemony that originary reproduction was eventually to extend over every aspect of everyday life in the latter part of this century. Tabloids, television, shopping malls, theme parks, video games, and computer-generated simulations of all kinds: all attest to the increasing domination of everyday experience by so many mass-produced simulacra that effectively undercut in advance any notion of a referent that would precede their reproduction.[7] Indeed, Baudrillard goes so far as to argue that the very predicate of reality has become increasingly dependent on the operation by which the "real" is reproduced: "The very definition of the real becomes: that of which it is possible to give an equivalent reproduction."[8] In short, it is no longer merely for the work of art, but for the whole domain of social practice that, in

Baudrillard's view, the notions of "original" and "originality" have ceased to bear the weight of any epistemic, political, or aesthetic authority.

Now, in the light of such an apocalyptic scenario, the question of the place of individual experience would seem to arise with particular urgency. What place could an individual subject occupy within this universe posited by Baudrillard, one increasingly dominated by operational principles and populated by industrial stereotypes? Baudrillard's response is that the effacement of the auratic quality of the original by the increasing dominance of serial reproduction leads to a decentering and emptying out of the individual subject in relation to the simulacra that precede it. In support of this argument, Baudrillard writes, for example:

> The simple presence of the television changes the rest of the habitat into a kind of archaic envelope, a vestige of human relations whose very survival remains perplexing. As soon as this scene is no longer haunted by its actors . . . , as soon as behavior is crystalized on certain screens and operational terminals, what's left appears only as a large useless body, deserted and condemned.[9]

Such is Baudrillard's argument: the subject withers away as a consequence of the death of the auratic object, persisting only as a condemned and useless vestige alongside the simulacra that precede and envelop it. Indeed, Baudrillard goes so far as to argue that the subject is not even permitted the reality of an authentic death, for our sacrifice to the simulation model—as in the strange reversal experienced by the spectators of Ballard's simulated crash—would take place precisely through our alienated resurrection in it. Such is Baudrillard's apocalyptic vision of contemporary culture: we are all sentenced not only to die, but to repeat ourselves posthumously, returning as our own cryogenic or hyperreal doubles.

> We . . . live in a universe everywhere strangely similar to the original—here things are duplicated by their own scenario. But this double does not mean, as in folklore, the imminence of death—they are already purged of death, and are even better than in life; more smiling, more authentic, in light of their model, like the faces in funeral parlors.[10]

Now, it goes without saying that this deliberately provocative and hyperbolic scenario invites objection on a number of counts.[11] Nonetheless, it seems to me that the appeal of Baudrillard's fashionable apocalyptic broadsides to a surprisingly broad public is of interest in itself. Baudrillard's work may be most usefully read as one articulation of a certain *phantasy* of postmodernity as a totalitarian operational system. Its interest would lie not in the truth or falsehood of its claims, but in the effects of truth it exerts on those who entertain and elaborate it. What, then, would be the stakes for an individual consumer who fervently embraced his or her consignment to vestigial status—and indeed to death—in relation to the mass-produced simulacrum? How are such apparent triumphs of planning as that described by Ballard—in which the anticipation of the most unforeseeable catastrophe is integrated into the production process—lived by the spectators in their pursuit of mute and solitary pleasures? How do they articulate what they take to be the originary simulation of their deaths with the rhythm of their condemned but obstinate existence? Above all, what is the relation of the simulacra invoked by Baudrillard to that "vestige" of human relations that seems to resist or exceed the scope of their operation?

Public Death and Private Phantasies

Ballard addresses these questions in his novel *Crash*, which may, from a certain perspective, be cited in support of Baudrillard. We have already seen the extent to which it lends itself to discussion in terms of Baudrillard's problematic of the simulation model: indeed, it is at one point, as we shall see below, discussed at length by Baudrillard.[12] But Ballard also opens the possibility of a critique of Baudrillard's arguments in their own terms, because it does not begin from the perspective of the "total system" that Baudrillard now vilifies, now celebrates, but never calls into question. On the contrary, Ballard's starting point is always that "mere vestige" that Baudrillard so scornfully dismisses—that of the fragmented and hollow subject, which nonetheless continues, from within an apparently operational universe, to long for its *outside*.

Ballard's black humor, as I shall show, turns on the distance between the consumer's ill-fated dream of breaking out of the interior and the world of the superior image that represents the interior's "outside" within it. His starting point is always the consumer's lair: the apartment in which husband and wife distractedly masturbate each other by the light of an evening newscast, "with all those scenes of pain and violence that illuminated the margins of our lives . . . , the beatings and burnings married in our minds to the delicious tremors of our erectile tissues" (p. 37); the cabin of the car, in which the driver dreams of reenacting the fatal collision of a starlet, joining "his mucous surfaces . . . to the wounds of this minor actress through the medium of his own motor-car (p. 189). Private desire, sealed in its interior, can seemingly circulate only in relation to the more rapid public image that figures its transcendence: and the image of the crash is the figure of the attempt to transcend the determinate space and limited speed of the subordinate everyday, and thereby to attain the absolute speed of the image on the far side of the screen. But it is the repeated failure of this attempt that generates the satirical and critical effect of Ballard's works: as we shall see, from this perspective, the "real" will appear not on the side of operationality and total planning, but on the side of the accident. It will appear, in other words, not as the product of the operation of the simulation model, but as that which resists it.

Before turning to these questions, however, it will be necessary to examine more precisely the extent to which Ballard's world coincides with what Baudrillard describes. In *Crash*, as in Baudrillard, the scenario of the subject's death appears first as at once a death sentence and a promise of resurrection. On the one hand, the spectator's anticipated death bears all the weight of mythic repetition in its mechanical and constraining aspect, appearing as a ritual or punitive reenactment of predestined scenarios. "Like everyone else bludgeoned by these billboard harangues and television films of imaginary accidents," writes Ballard's narrator, "I had felt a vague sense of unease that the gruesome climax of my life was being rehearsed years in advance, and would take place on some highway or road junction known only to the makers of these films" (p. 39). No Delphic oracle was more implacable or more feared than the pronouncements of the manipulators of actuarial statistics, in which Ballard's consumers have an implicit faith. But this submission to a predetermined fate has another aspect that, while no less mythic, compensates for the death sentence that the serial repetition of the crash has passed on its spectators by promising the fulfillment of a wish as old as humanity itself. These prospective martyrs of the civilization of the industrial accident can, through the very images of their anticipated deaths, lay claim to a peculiar sort of immortality that leads them to resign themselves to, and even to embrace, their violent fate. Once the long-awaited "original" accident finally takes place, launching them from the obscure privacy of their interiors into momentary stardom on the evening news, their death will reappear, eternally resurrected, in living rooms throughout the country. This is the phantasy that haunts Vaughn, the hero of *Crash*, who dreams of a fatal collision with Elizabeth Taylor that would launch him into a permanent afterlife on the far side of the screen.

Must one choose, then, between two alternative interpretations of the simulation model? Should we interpret the operational effect of the simulated crash as a compulsion imposed from without, or as the fulfillment of a long-held dream of resurrection? The beauty of the operational universe, as both Ballard and Baudrillard present it, is that it ultimately makes no difference. Whether embraced or feared by the spectators, the ultimate effect is the same: the simulacrum on the far side of the screen towers over the everyday, not as its representation, but as its mythic model and truer self. Judged by the simulation model, the spectators feel themselves to be inadequate to a role that they are obliged, and yet will never quite be able, to assume—that which has been predicted and displayed for them by stand-ins on the screen before it is played out by its original, if inferior, cast.

This sense of inadequacy of the everyday with regard to the simulation model gives rise to a curious passion for the real that, according to Baudrillard, dominates contemporary culture. *Crash* may be read as the tale of such a passion: it offers a staging of daily life's quest to attain the transcendent "reality" that glimmers just beyond the surface of the screen. Ballard's characters cling to Baudrillard's doctrine—"the very definition of the real [is] that of which it is possible to give an equivalent reproduction"[13]—with religious intensity. When Vaughn, a failed television personality, dreams of a fatal collision with Elizabeth Taylor that would fix ''his identity on some external event,'' the very nature of the ''event'' is inseparable from the mass production and circulation of its image (p. 168). Vaughn's own apartment is a veritable ''target gallery'' of screen actresses and public figures, a collage of movie stills, medical diagrams, and snapshots that serve to map out his planned trajectory toward his own violent apotheosis as he merges with the ''real'' image on the far side of the screen (p. 15).

Crash, however, is no simple celebration of this ''passion'' of which Baudrillard speaks. No doubt, in paying homage to ''the real'' as operationally defined, this passion at least implicitly legitimates the totality of the operational universe that gives rise to it; but in acknowledging the gap that separates the everyday from its superior counterpart, it also just as clearly gives the lie to the notion that the operational universe can command the totality of experience. When Vaughn ultimately encounters, in the comic collision with which the novel concludes, not the limousine of the modern Cleopatra, but the busload of package tourists who are borne happily along in her wake, the event underscores the unhappy consciousness of this consumer who, in aiming to derealize his body in the mythic space of spectacle, only succeeds in rejoining the other members of the audience. In his tale of Vaughn's obsession, Ballard thus shows the darker aspect of the anonymous consumer's right to fifteen minutes of fame. Ballard's victims, like those of Warhol's disaster series, live and die by the hierarchy that separates everyday life from its superior counterpart.

This inevitable comic failure of Ballard's characters—who end by falling short of their targets on the far side of the screen—suggests a definition of the ''real'' opposed to the operational one marketed by Baudrillard. From the perspective of everyday life, the ''real'' would no longer seem to be on the side of planning, but on the side of the accident; not on the side of operationality and performativity, but on that of malfunction and misfire: it would be that which resists and persists beyond the delimited space of the operation. From this perspective, the essential moment of the crash would be the violent rending open of the mobile but seemingly unbreachable interior of the car itself, which enacts a ''remaking of the commonplace'' ''as if intact memories and intimacies had been taken out of doors and arranged by a demolition squad'' (p. 52). In breaking open that interiority in which ''the intimate time and space of a single human being is fossilized forever'' (p. 12), the crash would appear as the emblem of an unarticulated desire to return to a ''primitive'' regime of ''symbolic exchange,'' to a relation of fluidity and ambivalence, between the domains of the living and the dead, and incompatible with operational criteria. Baudrillard himself celebrates *Crash* in these terms, as the articulation of the utopian desire for a body libidinally linked to the technological landscape with which it collides, oscillating in delicious ambivalence between death and desire. The shards of hurtling machinery signal through the flames to the wounds and scars that respond, not as substitutes for preexisting erogenous zones, but as "invaginations" that announce a new and savage eroticism,

> in the explosive vision of a body given over to "symbolic wounds," of a body merged with technology in its dimension of rape and violence, in the savage and continual operations it performs: incisions, scarifications, openings in the body (of which the "sexual" wound and its pleasure are only one example) . . . under the flashing sign of a sexuality with neither referent nor limits.[14]

Here the crash appears in retrospect, even to the immobile driver in the traffic jam, for whom the accumulated desire for movement has become its own obstacle: an "immense motionless pause" in which the unwilling "audience" of passengers appears to itself through its windows as already "resembl[ing] the rows of the dead" (p. 151). The wounds of his previous crashes, like a "bloody eucharist" (p. 157), continue to promise the utopian possibility of a desire that, no longer limited to inner private space, would take on a collective and metamorphic aspect:

> The silence continued. Here and there a driver shifted behind his steering wheel, trapped uncomfortably in the hot sunlight, and I had the sudden impression that the world had stopped. The wounds on my knees and chest [from previous crashes] were beacons tuned to a series of beckoning transmitters, carrying the signals unknown to myself, which would unlock this immense stasis and free these drivers for the real destinations set for their vehicles, the paradises of the electric highway. (p. 53)

Yet *Crash* is no more a simple celebration of or manifesto for the civilization of the industrial accident than a pious celebration of the simulacrum as a triumph of industrial planning. Ballard's consumers ultimately discover that the return to the happy (if violent) state of "savagery" invoked by Baudrillard is illusory, for if they experience the triumph of total planning as a regression to myth, here we see a contrary movement: the return to a primitivist's utopia of ambivalence and sacrifice will in the end turn out to be the product of a meticulously organized procedure. Ballard's Vaughn, in pursuit of an effect of disruption that mere chance can no longer be counted upon to produce, must devise a strategy every bit as elaborate as those of the engineers and marketing analysts whose power he flees. Engineering and medical studies, photographic records, and simulated accidents: Vaughn invokes them all in his pursuit of the transcendent Accident.

> Get all the paper you can, Ballard. Some of the stuff they give away—"Mechanisms of Occupant Ejection," "Tolerances of the Human Face in Crash Impacts. . . . " As the last of the engineers stood back from the test car Vaughn nodded appreciatively, and commented *sotta voce*,

> "The technology of accident simulation at the R.R.L. is remarkably advanced. Using this set-up they could duplicate the Mansfield and Camus crashes—even Kennedy's—indefinitely." (p. 123)

The operational universe and that of symbolic exchange thus pass into one another through the contradictory figure of the simulated crash, suspended between two seemingly incompatible definitions of the "real," each appearing as the illusory but necessary truth of the other, with neither ultimately able to prevail. On the one hand, the crash represents the triumph of the operational universe, its extension even to the point of engineering accidental death; but on the other, it appears as what is at the limit of that universe, as what is still capable of producing an effect of chance, even if this effect results increasingly from elaborate preparations. If the "real" is indeed lived, as Baudrillard observes, in operational terms as "that of which it is possible to give a mechanical reproduction," it is at the same time lived as what withdraws absolutely from its reproduction, as the absolutely unique Catastrophe whose absence stands in for the vanished auratic object.

This is clearest in Vaughn's "real-life encounter" with his intended victim. Shielded by an impenetrable and ubiquitous hymen of tabloids and publicity photographs, the distance separating the film star from an audience whose imagination is itself a screen—"a target gallery of screen-actresses, politicians, business-tycoons and television executives" (p. 15)—is reduced to a minimum but remains absolute. The target's protection is the gallery itself:

> The isolated figure of the screen actress stood beside her chauffeur, a hand raised to her neck, as if shielding herself from the image of the death she had so narrowly avoided. The police and ambulance men, the crush of spectators squeezing themselves between the parked police cars and ambulances, were careful to leave a clear space around her. (p. 222)

Vaughn himself will thus discover no "outside" on the far side of the screen, at least not in the sense it was imagined. There is no real other than the relation to the spectacle he has shared all along with his fellow victims—the other members of the audience. The crash is not the copula equating private disaster with public memory. On the contrary, it only succeeds in displacing the barrier it attempts to transgress and in reenacting the contradiction it seeks to resolve. The narrator's first crash serves in this sense as the prototype for all the rest. The desires of two anonymous drivers give rise to the phantasy of a literally explosive encounter that would expel the body from the private domain, submitting it to the risk of what appears to be a chance operation; and yet, in the end this accident only succeeds in repeating the situation they were meant to escape:

> The same mysterious forces that saved me from being impaled on the steering wheel also saved the young engineer's wife. . . . All I could

> see in my mind was the image of the two of us locked together face to face in these two cars, the body of her dying husband lying between us on the bonnet of my car. We looked at each other through the fractured windshield, neither able to move. Her husband's hand, no more than a few inches from me, lay palm upwards beside the right windshield wiper. His hand had struck some rigid object as he was hurled from his seat, and the pattern of a sign formed itself as I sat there, pumped up by his dying circulation into a huge blood-blister—the triton signature of my radiator emblem.
>
> Supported by her diagonal seat belt, his wife sat behind her steering wheel, staring at me in a curiously formal way, as if unsure what had brought us together. Her handsome face . . . had the unresponsive look of a madonna . . . , unwilling to accept the miracle or nightmare sprung from her loins. . . . Did she realize that the blood covering my face and chest was her husband's? (p. 21)

If one adheres to Baudrillard's reading of Ballard, the corpse of the dead husband, expelled from the interior, would be the uncanny object of symbolic exchange, transmitting an ambivalent charge of aggression and desire, conducting a flow of blood and semen across its limits; and, indeed, a phantasy of something like "symbolic exchange" undoubtedly informs the narrator's initial experience of the crash. Nonetheless, all strangely returns to the image behind the screen, to the victim-become-icon: the madonna. The limit separating the two participants in this attempted exchange is maintained: each remains a spectator of the other's mutilation. "We looked at each other through the fractured windshields, neither of us able to move" (p. 20). The barrier that seals the subject of desire in its isolation is reduced to its minimum, but this minimum is sufficient; the shattered windshield preserves the wounded spectator like a protective shroud. Desire, as it nears the limits of the screen, recoils, expelling the sacrificial substitute by which it hopes to mime and communicate with its own death; it ultimately succeeds only in displacing the limit it had hoped to transgress. Between the wounded spectators, the corpse of the husband attests to a sacrificial logic that is beyond them: already a mythologized corpse, bearing the stigmata of the trademark, he already belongs to an order that they can only contemplate from a distance. The unpredictable but willed scene of disaster can be imagined only as spectacle. What exceeds the spectacle, on the far side of the screen, is not what is outside it, but what is imagined in its place.

The Simulation Model and Its "Outside"

In his dreams of transcendent union with the electronic image, Vaughn thus proves to have been all along merely the figure of a public that invests such spectacles with a waning auratic power. Embracing the death sentence that has been passed upon it with sacred fervor, he merely takes the unspoken dreams of that public—in both its aspect as spectator (figured in the visitors to the Road Research Laboratory) and its aspect as sacrificial participant (figured in the tourists with whom he ultimately collides)—to their logical conclusion. At some level, we are all like Vaughn, for, as with the player of a video game who "dies" at the end of each scenario, only in order to find him- or herself once again resurrected (after the deposit of another coin) to operate the controls, the content of our death never ceases to return to us, in violent or apocalyptic reconstructions from which "we" are necessarily absent, basking in the posthumous glow it bestows upon us. But the form of the consuming subject as at once witness and prospective victim of such fatal scenarios is maintained, even as its content is negated: this is, for the most part, how "the death of the subject" is staged in late capitalist culture, however subversive it might appear in the abstract.

It is for this reason that the utopian yearning linked to the crash ultimately finds its truest expression, not so much in the phantasmagoria of the primitive, with its ritual flames and savage scars, as in the comic failure of the model itself, which confirms the gap between the simulated scenario and the debased "original cast" destined to act it out. In the path that should have led him to his terminal collision with the movie actress, Vaughn's trajectory is, as we have seen, interrupted by his encounter with a public (the busload of tourists) that appears outside the scenario prefigured by the simulation model. Here, the technology of simulation, which had seemingly drawn death and desire alike into the domain of rational planning, is derailed in its encounter with this shapeless crowd, the ultimate destination of Vaughn's desire. This collective is not, however, imagined as a regression to a community preceding the technology of simulation: on the contrary, the everyday world from which it emerges appears only as the outside *of* the order of simulacra itself—as its inferior and incomplete image. But it is this very incompleteness that offers Vaughn a measure of redemption in the end, as the crowd enshrouds him in its stubborn anonymity. In falling from the mythic domain of the simulation model into the embrace of the dreaming collective, his identity is comically transfigured and refunctioned. Vaughn's fate thus attests to the nonidentity of the simulation model across its repetitions and appropriations, even for the subject who imagines its dominion to be totalitarian and absolute; and it is thus that Ballard clears from beneath the wreckage of the simulation model a possible site for utopian thought.[15]

Notes

1. Siegfried Giedion, "Mechanization and Death: Meat," in *Mechanization Takes Command: A Contribution to Anonymous History* (New York: Norton, 1969), 209-46. Page numbers appear in text for further quotations from this work.

2. "A Klee painting named 'Angelus Novus' shows an angel looking as though he is about to move away from something he is fixedly contemplating. His eyes are staring, his mouth is open, his

wings are spread. This is how one pictures the angel of history. His face is turned toward the past. Where we perceive a chain of events, he sees one single catastrophe which keeps piling wreckage upon wreckage and hurls it in front of his feet. The angel would like to stay, awaken the dead, and make whole what has been smashed. But a storm is blowing from Paradise; it has got caught in his wings with such violence that the angel can no longer close them. This storm irresistibly propels him into the future to which his back is turned, while the pile of debris before him grows skyward. This storm is what we call progress.'' Walter Benjamin, *Illuminations*, ed. Hannah Arendt, trans. Harry Zohn (New York: Schocken, 1969), 257-58.

3. For an account of these impulses and their antinomies in the quintessentially modernist poet, see ''On Some Motifs in Baudelaire,'' in ibid., 155-200.

4. J. G. Ballard, *Crash* (New York: Vintage, 1985), 126-27. Page numbers appear in text for further quotations from this work.

5. Jean Baudrillard, *Simulations*, trans. Paul Foss, Paul Patton, and Philip Bechtman (New York: Semiotext[e], 1983), 50.

6. Benjamin, *Illuminations*, 217-51.

7. ''We know that now it is on the level of reproduction (fashion, media, publicity, information and communication networks) . . . , that is to say in the sphere of simulacra and of the code, that the global process of capital is founded.'' Baudrillard, *Simulations*, 99.

8. Ibid., 146.

9. Jean Baudrillard, ''The Ecstasy of Communication,'' in *The Anti-Aesthetic: Essays on Postmodern Culture*, ed. Hal Foster (Port Townsend, Wash.: Bay, 1983), 129.

10. Baudrillard, *Simulations*, 23. The process observed by Walter Benjamin would now seem to have reached full fruition: with the expropriation of ''my death'' the last vestige of the auratic has been exorcised from everyday experience, to reappear on our screens under the sign of aesthetic disinterestedness. ''Mankind,'' writes Benjamin, ''which in Homer's time was an object of contemplation for the Olympian gods, now is one for itself. Its self-alienation has reached such a degree that it can experience its own destruction as an aesthetic pleasure of the first order.'' Benjamin, *Illuminations*, 242.

11. One might readily concede, for example, that the increasing interweaving of our lives with technically manipulated and reproduced images has reached a point that we are less and less willing to lay claim to an original reality that would precede the ubiquitous simulations through which we pass. Indeed, Jameson has convincingly argued that our capacity as individual or collective subjects to bind the content of an everyday experience increasingly fragmented into noncommunicating subsystems is decidedly on the wane: ''The subject has lost its capacity to extend its pro-tensions and re-tensions across the temporal manifold, and to organize its past and future into a coherent experience.'' Fredric Jameson, *Postmodernism; or, The Cultural Logic of Late Capitalism* (Durham, N.C.: Duke University Press, 1991), 25.

Nonetheless, the persistence alongside these tendencies of practices long associated with the much-maligned centered and humanist subject—which would presumably extend to writing elaborately reasoned and provocative books such as Baudrillard's—should at the very least give any reader pause. As Eagleton has argued, ''The bourgeois subject is not in fact simply part of a clapped-out history we can all agreeably or reluctantly leave behind: if it is an increasingly inappropriate model at certain levels of subjecthood, it remains a potently relevant one at others. Consider for example the condition of being a father and a consumer simultaneously. The former role is governed by ideological imperatives of agency, duty, autonomy, authority . . . [which are generally associated with the unified individualist subject]; the latter, while not wholly free of such strictures, puts them into significant question. . . . The two roles are not of course merely disjunct; but though relations between them are practically negotiable, capitalism's ideal consumer is strictly incompatible with its current ideal parent.'' Terry Eagleton, ''Capitalism, Modernism and Postmodernism,'' *New Left Review* 152 (July/August 1985), 71-72.

Eagleton argues, in other words, that the tendency toward the subject's dissolution is, whether for good or for ill, contradicted by other, countervailing tendencies that persist in contemporary culture. To return to Baudrillard's example, television has not yet managed to eradicate the actors completely from the private scene of which it is only a part; leaving aside the moral dilemmas of paternity to which Eagleton alludes, this is perhaps most clearly the case with the repressive exercise of paternal power, which is by no means confined to simulation models and subtle disciplinary procedures. Any child who has faced the tyrannical menace of a father's belt would question whether the television's blare could be capable of completely drowning out the sound of its blows, not to mention whether any number of video games could completely dispel the old oedipal culture with which they resonate.

As Eagleton puts it, ''Many subjects live more and more at the points of contradictory intersection'' (p. 72) between these various tendencies in late capitalist culture, and while one should no doubt applaud Baudrillard's observations that the ''deconstruction of the subject'' is not merely the creature of exotic philosophical speculations, but is on the contrary inseparable from the experience of everyday life, one must at the same time agree with Eagleton that it remains only a tendency of contemporary culture, one that may be expected to be unevenly developed from one domain of social praxis to another.

12. See Jean Baudrillard, ''Crash,'' in *Simulacres et simulations* (Paris: Editions Galilée, 1981), 165-78.

13. Baudrillard, *Simulations*, 146.

14. Baudrillard, *Simulacres et simulations*, 166; my translation.

15. For an elaboration of the utopian uses of the simulacrum, see Scott Durham, ''The Poetics of Simulation: The Simulacrum and Narrative in the Works of Jean Genet and Pierre Klossowski,'' doctoral dissertation, Yale Univerity, 1992.

PART IV
Technology and Cyberspace

Chapter 10

The Seductions of Cyberspace

N. Katherine Hayles

Technology is literary criticism carried on by other means.
—Bruno Latour

Hans Moravec has a dream. A roboticist at Carnegie-Mellon University, Moravec wants to download the information stored in the human brain and transfer it to a computer. In his view information is information, whether stored in silicon-based hardware, disk software, or cranial wetware. Once the transfer is complete, the body becomes disposable, an outmoded artifact to be discarded along with the limitations of space and time that it necessitated. Moravec is not crazy; he is head of Carnegie-Mellon's Mobile Robot Laboratory. And he is not alone. His dream is shared by many others, appearing with variations in fields as diverse as cryogenics, genetic engineering, and nanotechnology.[1] Ed Regis has identified this dream as the "desire for perfect knowledge and total power. The goal [is] complete omnipotence: the power to remake humanity, earth, the universe at large. If you're tired of the ills of the flesh, then *get rid of the flesh*; we can *do* that now."[2]

Perhaps not since the Middle Ages has the fantasy of leaving the body behind been so widely dispersed through the population, and never has it been so strongly linked with existing technologies. The conjunction with technology is crucial. In its contemporary formulation, the point is not merely to leave the body but to reconstitute it as a technical object under human control. The essential transformation is from biomorphism to technomorphism. The transformation has important implications for every area of contemporary culture, including literature and literary criticism. It is not for nothing that we speak of the body of a text and the corpus of literature. Our sense of our physical bodies, their capabilities and limitations, boundaries and extensions, deeply informs both the objects and the codes of representation. Less clear are the implications of these mappings. In this last decade of the twentieth century, elisions between physical and textual bodies are entangled with complex mediations that merge actual and virtual realities, ideological and technological constructions.

The issues are joined in the emerging technologies of cyberspace (also called virtual reality, VR, and artificial reality). Assuming various forms, these technologies splice a human subject into a cybernetic circuit by putting the human sensorium in a direct feedback loop with computer data banks. VR breaks the barrier of the screen, opening the high-dimensional space beyond to sensory as well as cognitive habitation by the user. With VR you don't just *see* data banks; you can sit down on them and watch the river of information flow by. Or you can plunge into the river. Turning Heraclitus on his head, Michel Serres has asserted that flows are more constant than the material world that expresses and embodies them.[3] The river remains the same, while the banks constantly erode and change. Body cells change and die; it is the flow of energy and information through the organism that maintains continuity. No man steps twice into the same river not because the river changes, but because he does. These inversions are consistent with virtual reality, for they figure the flow of information within systems as more determinative of identity than the materiality of physical structures. Plunging into the river of information implies recognizing that you *are* the river.

Baudrillard has written about the implosion of cultural space that takes place when the copy no longer refers to an original but only to another copy.[4] Defining a simulacrum as a copy with no original, Baudrillard imagines a precession of simulacra (*precession* is a mathematical term denoting the gyration of a sphere when spinning under torque, as when a top slows down and begins to wobble). The spinning metaphor is appropriate, for in the circular dynamic in which copy replaces copy until all vestige of the original is lost, reference is supplanted by reflexivity. Virtual reality exemplifies the implosion Baudrillard describes. When the technologically enhanced body is joined in a sensory feedback loop with the simulacrum that lives in RAM, it is impossible to locate an originary source for experience and sensation. The "natural" body, unmodified by technology, is displaced by a cybernetic construct that consists of body-plus-equipment-plus-computer-plus-simulation.

Within the cultural space that VR occupies, the arrows of signification do not all point the same way. The double hermeneutic of suspicion and revelation that Fredric Jameson advocates is appropriate to interrogate its multiple significances.[5] The drive for control that was a founding impulse for cybernetics (defined by Norbert Wiener as the science of control and communication) is evident in the simulations of virtual reality, where human senses are projected into a computer domain whose underlying binary/logical structure defines the parameters within which action evolves. At the same time, by denaturalizing assumptions about physicality and embodiment, cybernetic technologies also contribute to liberatory projects that seek to bring traditional dichotomies and hierarchies into question. Ironically asserting, "I would rather be a cyborg than a goddess," Donna Haraway sees the cyborg as offering feminists a metaphor that cuts through the Gordian knot tying woman together with nature, thereby freeing us from the burdens that conjunction imposes.[6]

There are, however, new burdens imposed by constructing woman (and man) as cyborg. The turn is characteristic of virtual reality, for the space within which it operates is intensely ambiguous. For every solution it offers, it raises new problems; for every threat that erupts, new potentialities also arise. Countering the fetishistic drive for control is the spontaneous, free-flowing collectivity that emerges when multiple players in virtual reality collaborate to build a world. Offsetting the creation of technosubjects is VR's ability to leapfrog over abstraction, returning to the reconstituted subject the rich diversity of a sensorium that includes visual, kinesthetic, and tactile experience. Compensating for the underlying machine logic that, for all its versatility, is the Procrustean bed into which human perception must fit is the thrill of creating and exploring virtual worlds.

The point is not to resolve these ambiguities—a quixotic adventure, since they will not yield to theoretical pronouncements alone—but to use them to understand the cultural forces driving the technologies forward and determining how they will be used. As Bill Nichols argues, we should ask "what tools are at our disposal and what conceptions of the human do we adhere to that can call into question the reification, the commodification, the patterns of mastery and control" that are simultaneously reinforced and exposed by these technologies.[7] To the extent they are reinforced, the patterns are more difficult to break; to the extent they are exposed, they become subject to analysis and therefore to change. Moreover, the technologies themselves can be—already are—agents of change. This is the double edge of virtual reality's revolutionary potential: to expose the presuppositions underlying the social formations of late capitalism and to open new fields of play where the dynamics have not yet rigidified and new kinds of moves are possible. Understanding these moves and their significances is crucial to realizing the technology's constructive potential.

Full-Body Processing

The technological development of cyberspace began, as did so much else, in the 1960s. As early as 1968, Ivan E. Sutherland at the University of Utah had the idea of creating a head-mounted display that connected a user directly to a computer. The device was so heavy it had to be suspended from the ceiling, but its possibilities were enticing. Other lines of development ran through Myron Krueger, who did a dissertation on artificial reality in the late 1960s at the University of Wisconsin, and Fred Brooks at the University of North Carolina.[8] Krueger's vision differed from Sutherland's because he wanted participants to be able to move freely, unhampered by heavy equipment. His approach used sensing devices to determine a participant's position and body movements, which were then fed into a computer to create interactive graphic displays. By 1985 the distance between the two approaches had diminished considerably. The technology was available to miniaturize the head display, making it a portable helmet rather than a dangling behemoth. By then military and government agencies had picked up on the idea. Convinced of the potential, NASA earmarked several million dollars for cyberspace projects. The U.S. Air Force budgeted a similar amount for its ongoing Super Cockpit project, which uses virtual reality simulations to direct the pilot's interactions with the aircraft. Video games provided models for the simulation programs. When William Gibson coined the term "cyberspace" in *Neuromancer* (1984), the novel that sparked the cyberpunk literary movement, he was working from a sense he had gotten from video game freaks that a space existed behind the computer screen that was as interesting as, or more interesting than, the space in front of it.[9]

The technology took a quantum leap forward in the late 1980s. Stimulated by reading *Neuromancer*, John Walker of Autodesk, a software company specializing in computer-assisted design (CAD) packages, issued a white paper calling for a major investment in cyberspace software. Arguing that the screen was the next barrier to be broken, Walker defined a cyberspace system as "a three-dimensional domain in which cybernetic feedback and control occur."[10] Somewhat earlier, VPL had begun to take off, a company devoted to virtual reality technologies and headed by Jaron Lanier, the dreadlocked guru of VR, who also is a shrewd businessman. VPL found a market for its products in the video game business, designing the PowerGlove for Nintendo. The company also developed virtual reality software and paraphernalia, including a stereo vision helmet and the DataGlove, a more sophisticated and interactive version of the PowerGlove.

The idea behind the technologies is to create a feedback loop between the user's sensory system and the cyberspace domain, using real-time interactions between physical and virtual bodies. In one version, the player's movements and reactions are monitored through such input devices as stereo-vision helmets and data gloves. Flex your fingers in VPL's DataGlove and the simulacrum representing you in cyberspace moves to pick up the object you see in the helmet's monitor. Glance around and the virtual perspective changes accordingly, creating with a slight time lag the scene you see in the helmet's stereovisual field. Turn the bars of your cyberbike and the puppet's bike zooms in a different direction. Alternatively, you may turn your bars to avoid the car that comes whizzing toward the puppet. Stimuli go in both directions; what happens to the puppet has an impact on your sensory field, just as what you do affects the puppet. The puppet is a version of and a container for the self. It is, as Randall Walser, a senior programmer at Autodesk, writes, "a vehicle for your mind. Looking through the puppet's eyes, your sense of self merges with it, so that . . . you are the puppet and the puppet is you."[11]

One advantage of cyberspace over ordinary reality is its flexibility. Puppets may be directed by artificial as well as human intelligences, creating a three-dimensional field of play in which silicon- and protein-based life forms interact. It is also possible to switch one's viewpoint between puppets or invest it in a "spirit," a disembodied space that represents the point from which the user interacts with the cyberspace environment. In VPL's "Reality Built for Two," a game of cybertag, one strategy is to hide in the other player's head. Potential users of virtual reality include architects, who can stroll around the inside of buildings before they are built; astronauts, who can use the cyberspace puppets to direct robots outside spacecraft; and fitness club instructors, who can interface exercise equipment with cyberspace to create adventures that will spice up their patrons' exercise routines.

Cyberspace can also be used to cope with that affliction of the postmodern age, *too much information*. Creating direct feedback loops between data and human senses allows information to be processed holistically, much as environmental cues are. Michael Spring asks us to imagine entering a virtual reality library, forming a research question, and watching as colors and configurations change in response to the question.[12] Corroborating evidence appears in hot colors, contrary facts in cool. Lines appear linking data formations and indicating their relationships to one another. Data directly relevant to the question are connected by heavy dark lines, secondary data by broken lines. As another question is asked, or the first question rephrased, colors and configurations change accordingly. The idea, Spring notes, is to "suggest visual metaphors for mental models of how the idea space is organized."[13] A similar proposal made by Scott Fisher would enable a user wearing a helmet and bodysuit to touch a screen and arrange blocks of data in a projected three-dimensional space. Direct experience, Fisher writes, "has the advantage of coming through the totality of our internal processes—conscious, unconscious, visceral and mental—and is most completely tested and evaluated by our nature."[14] VR allows the user to draw on that totality in ways that most information-processing systems do not.

Spring's model emphasizes vision and Fisher's kinesthesia, but the reasoning behind them is the same. Why throw away the advantages bestowed by millennia of evolution to dwell in the realms of abstract concepts when we have the capability to use full-body processing? In the collaboration that virtual reality sets up between the human sensorium and computer memory, the sophisticated and nuanced response to environmental cues that has enabled human beings to dominate the planet is joined with the power of computers to store, process, and display information. It represents, some would say, the best of both worlds—or perhaps the next leap forward in technobioevolution. From the protein-based life form come the flexibility and sophistication of a highly complex analogical processor that includes sensory, unconscious, and conscious components; from the silicon-based entity come massive storage and combinatorial ability, rapid retrieval, and reliable replication.

That the subjectivity that emerges from this joining is a cyborg rather than a human can scarcely be missed, although neither Fisher nor Spring comments on the fact. Already about 10 percent of the U.S. population are cyborgs in the technical sense, including people with electronic pacemakers, prosthetic limbs, hearing aids, drug implants, and artificial joints. VR would substantially increase this percentage. If the extent to which one has become a cyborg is measured in terms of impact on psychic/sensory organization rather than difficulty of detaching parts, VR users—cybernauts, some writers prefer to call them—are more thoroughly cyborgs than are people with pacemakers. The reorganization of subjectivity that VR effects is not, of course, limited only to this technology. As William Gibson noticed several years ago, video game players and word processing users are also spliced into cybernetic circuits with their machines, with resulting reorganization of their neural networks.[15] VR extends rather than initiates this reorganization, making explicit transformations that have been under way for some time.

Imagine walking into a virtual reality library and asking, "How many of the human populations of the planet are cyborgs?" A hologram of the earth appears before you, with hot colors indicating areas of high density, cool colors indicating relatively unmodified humans (your own suit, of course, is colored very hot). Now ask, "Which of the human populations on the planet are absorbing more than their fair share of the planet's resources?" Would the hologram change? The scenario implies that issues of class, race, and gender are likely to be replayed in a different key, in which the mark of privilege is access to cyborg modifications. In a time of rapid realignments in cultural formations, when the populations of the planet are extremely heterogeneous with respect to the coming changes, questions of how cyborgs relate to unmodified humans will be central.

A window onto these issues is opened by Joseph Henderson, a physician associated with the Interactive Media Laboratory at Dartmouth Medical School. Programs already exist that make use of VR for medical purposes. In Electronic Cadaver, VR interactions are used to simulate dissection, so that medical students can move scapels and get appropriate kinesthetic and visual feedback without the necessity of formaldehyded bodies. Henderson describes Traumabase, another VR medical training program. Traumabase works with a multimedia data base generated during the Vietnam War on medical casualties, including 200,000 sheets of paper, 50,000 slides, hours of audio recordings and film, and the videodisc history *Vietnam: The 10,000 Day War*. Henderson points out the difficulties inherent in accessing this much information; conventional programs do not allow users to "interact with data and information in the same way we think, moving rapidly and linking item to item, idea to idea, analysis to analysis."[16] Traumabase uses VR techniques to create an information matrix that can be "explored and navigated'' to reveal ''expected and unexpected patterns'' of ''location and severity of wounds, wound pattern clustering, wound pattern frequencies, survival patterns.'' An ''interactive process of discovery can result'' that uses ''the very powerful combination of eye, brain, and hand. This can provide a 'visceral' sense or analysis of what the data have to tell us.''[17]

Henderson links this ''visceral'' processing with a more fully human reaction to what the data represent, contrasting it with traditional scientific analyses:

> In the interest of ''rational'' or ''scientific'' decision-making we isolate the quantifiable and formulate models. A danger is that the abstraction can become the reality, and real world decisions can be made without due regard to the real world. However, in this system abstractions (the matrix) can be linked to increasingly concrete and emotionally powerful forms of information, to the realities of seeing people and hearing their stories. With this kind of approach we can involve the heart as well as the mind.[18]

The dichotomy between abstract analysis and the ''real world'' that the passage constructs elides the difference between actual and virtual realities. Seeing a cybernetically reconstructed body and hearing a voice recording slides into seeing a wounded man and hearing him scream. The elision is not trivial. Granted that the VR reconstruction is laden with more sensory information than statistics, there is still a chasm separating the virtual simulation and the physical reality of mangled bodies.

The complexity of the issues precludes simple resolution. Henderson is certainly correct in contrasting statistical abstraction with the VR simulation's greater emotional impact. Underlying this contrast, however, is the reconstruction of subjectivity that VR implies. Being able to occupy a virtual space implies that one can have the benefits of physicality without being bound by its limitations. One of the most emotionally charged of these limitations is mutilation or death of the physical body. The privileged position that virtual reality bestows upon the subject marks a difference between him or her and others who cannot enter this space, specifically those wounded or killed in the war. Their simulacra enter the virtual space only to testify to their inability to reconstitute themselves as virtual subjects removed from the perils of physicality. The very sensory stimulation that Henderson sees as constituting an empathic bond between victim and user reinstitutes difference in another register. The Traumabase user may not, of course, consciously recognize this difference. Its effect would be even more powerful if registered below the level of conscious awareness.

Eros and the Cyborg

The problematic relations between sense and empathy, virtual user and physical object, hint at how psychic and social life may be reorganized when virtual reality comes into widespread use. The possibilities are as diverse as the human imagination. One scenario imagines virtual parties, where the participants never meet face-to-face but interact through their cyberspace surrogates. Randal Walser writes that the cyberspace user, "unconstrained by physical space," will begin "to work, play, learn and exercise in magical new worlds."[19] The essence of this "magic" is the construction of the body as an *absent signifier*. After visiting VPL, John Perry Barlow reported, "It's like having your everything amputated."[20]

Nowhere are the problematic effects of VR clearer than in the realm of the erotic. Barlow remarks that he has been through "eight or ten Q. & A. sessions on Virtual Reality and I don't remember one where sex didn't come up." "This is strange," he muses. "I don't know what to make of it, since, as things stand right now, nothing could be more disembodied or insensate than the experience of cyberspace."[21] In another sense, the evocation of the erotic is anything but strange. Bruce Clarke has pointed out that the violation or dissolution of body boundaries is inherently erotic; the same observation has been made by writers as diverse as Ovid, Saint Teresa, and the Marquis de Sade.[22]

The juxtaposition of eroticism, violated taboos, and modified bodies helps to explain why the high-tech world of the cyborg should so frequently take on a Gothic tinge. In Vernon Vinge's "True Names" (1981), often identified as the original cyberspace story, castles and dragons populate the landscape of the Other Planet, a consensual space created when humans strap on electrodes to interface with each other and artificial intelligences through computer networks.[23] Once on the Other Planet, the user's consciousness is manifested through whatever form he or she desires. One appears as a beautiful red-haired woman; another as a typewriter. The Gothic landscape and creatures that surround these forms are more than quaint anachronisms. Rather, they serve as tropes that map complex cultural formations onto the technomorphisms unique to the twentieth century.

We can trace the mapping by considering the mingling of magic and technomorphism signified by the title. In a preface, Vinge explains that he thought of "True Names" after reading Ursula LeGuin's Earthsea trilogy.[24] Central to LeGuin's trilogy is the belief, common to magical traditions from fairy tales to voodoo, that knowing someone's true name gives one power over that person. In Vinge's narrative, knowing someone's true name means discovering that person's prosaic everyday identity, along with his or her social security number and, most important, home address. Whereas in LeGuin the true name's power derives from the conflation of signified with signifier, in Vinge it comes from being able to locate the physical body, with all of its frailties and vulnerabilities, from which consciousness emanates. The conflation here is not of name and thing, but of biomorphism and technomorphism.

The Gothic allusions reinforce a homology also constructed through action and plot: as signifier is to signified, biomorph is to technomorph. Through the homology the body becomes a gesture pointing toward the "real thing" rather than the thing itself. The reality is the technomorph, the body an atavistic vestige that functions as an Achilles heel, limiting the technomorph's power. It comes as no surprise at the story's end when one of the characters chooses to transfer her mind into a computer. Shedding the Achilles heel of her physicality bestows immortality upon her. It also allows her to assume the privileged role of guardian to humankind's impending transformation into technomorphs. It is a dream Hans Moravec would recognize—and not only a dream: increasingly, a technology as well.

In the same issue of *Mondo 2000* as Barlow's puzzlement over why VR and eroticism should so often go together, Howard Rheingold has an article that reveals how powerfully the absence of physicality can interact with eroticism to form fantasies deeply characteristic of our cultural moment. Rheingold envisions a technology that he calls "teledildonics"; he writes:

> Before you climb into a suitably padded chamber and put on your headmounted display, you slip into a lightweight—eventually, one would hope diaphanous—bodysuit. It would be something like a body stocking, but with all the intimate snugness of a condom. Embedded in the inner surface of the suit, using a technology that does not yet exist, is an array of intelligent effectors. These effectors are ultra-tiny vibrators of varying degrees of hardness, hundreds of them per square inch, that can receive and transmit a realistic sense of tactile presence in the same way the visual and audio displays transmit a realistic sense of visual and auditory presence.[25]

The idea is to plug the bodysuit into a telephone that has a visual screen on which you and your communicant are displayed. The information coming over the telephone interacts with the bodysuit effectors to provide kinesthetic and tactile sensations appropriate to the visual and audio messages. The result, Rheingold intimates, is the ultimate safe sex.

The teledildonic fantasy illustrates how the body as absent signifier plays into the eroticism of metamorphosis. The body is transformed into a technomorphism not only through the visual display but also through the kinesthetic sensations that reinforce, in a different sensory loop, the audio transmissions. The metamorphosis is not into a different biological form but into the cyborg that results from splicing together the physical and virtual bodies. The cybernetic long-distance coupling between communicants replays on a different level the reconstitution of body boundaries that has already taken place through the technology. Further reinforcing the fantasy, and close to the surface, is a strong anxiety about the perils of physicality, especially AIDS. Add to this the growing suspicion among the population that time-release environmental poisons are making physicality an impractical state to inhabit, and the appeal of virtual reality is obvious.

So, too, are its dangers. Establishing a dialectic between actual and virtual objects, VR invites a hierarchy to be set up between them. If we can believe what our writers are telling us, the vectors will run from virtual to actual, privileging computer construct over physical body. Virtual reality is not the only factor determining this order. Also contributing are other technologies that make the body into a commodity, from organ transplant depositories (significantly called "banks") to cosmetic surgeries. As the body increasingly is constructed as a commodity to be managed, designed, and parceled out to deserving recipients, pressure builds to displace identity into entities that are more flexible, easier to design, less troublesome to maintain.[26]

Gibson's *Neuromancer* illustrates how the technologies of informatics and body management come together to create a world where the virtual body is the "real thing," the physical body a mere substitute. This is a world of fast burnout, generation gaps between sixteen- and twenty-year-olds, investment in styles that change overnight. Styles are, moreover, expressed not only through clothing, but also through designer drugs, facial and full-body surgery, cybernetic splices into the human neurosystem of computer chips, and various other kinds of sensory interfaces. Commercial products, mentioned by name, are scattered all over the surface of this text, from high-tech computers to the latest body modifications. The same impatience shown toward an outmoded computer is directed toward unreconstructed bodies, from the protagonist's disdain of his own body to the women who are programmed through computer interfaces to act as prostitutes while their minds are parked elsewhere. They are called, significantly, "meat puppets."

Neuromancer enunciates a new axis along which wealth and power will operate, as they already operate along the axes of gender, race, and class. Behold the axis of physicality. The privileged end is the virtual, the stigmatized end the physical. Having an unmodified body will be like having a working-class accent; it will mark you as cannon fodder for the system. Body politics, already well articulated within feminist theory, will mean not only the imbrication of the body in gendered structures, but also a politics of physicality shaped by the technologies of technomorphism and informatics, including computer simulations, cybernetics, genetic engineering, organ transplants, bioactive drugs, and reconstructive surgery.

Although the terrain on which these struggles will take place is largely unmapped, some of the possibilities have been envisioned in contemporary fiction. Tom DeHaven's *Freaks Amour*, an underground classic, records the struggles of mutants to become "norms."[27] Victims of the fallout from a mysterious radioactive blast in New Jersey, they dream of synthetic skin and full-body surgery that will restore them to invisibility and social acceptance. At least some do. Others argue that freakishness ought to be embraced, worn as a badge of honor in the fight against the power structures responsible for the blast—which, it turns out, are the same forces who plan to co-opt the hallucinogenic mutigens the blast has created to escape from the planet. First they use technology to poison the planet, then they develop it further to escape from the planet they have poisoned.

The reasoning reveals why body politics is at the center of contestations for power at the century's end. Only a small minority—if indeed any—of the planet's population will be able to escape from the state of physicality that most of us will continue to inhabit. The fantasy that escape is possible authorizes people to believe that they will be among the chosen few, that we will not have to continue to live with the messes we have created. Since cleaning up those messes may be impossible (how will we repair the damage to the ozone layer?), the need to believe that escape is possible is very strong. To the extent that cyberspace plays into this fantasy, it contributes to a continuing unwillingness to face problems that are not going to go away. In some contexts, leaving the body behind equates to the belief that if the problems won't go away from us, perhaps we can go away from the problems. Is it necessary to insist that nothing could be further from the truth?

The Body Zone

Marked bodies, the longing for invisibility, stigmata that also become sources of strength—the themes are familiar, running from Ellison's *Invisible Man* to Philip K. Dick's *The Three Stigmata of Palmer Eldritch* to Katherine Dunn's *Geek Love*. The continuities suggest how the new technologies will extend and complicate body politics, as well as how dynamics already in play will be mapped onto the simulated grounds of virtual reality. Many of these dynamics concern gender. It is no accident that the protagonist of *Freaks Amour* plans to get the money for reconstructive surgery by putting on a freak show in which he rapes his sweetheart.

The next time you are in a shopping mall, check out the video arcade. Most of the patrons are teens and preteens. How many are male? If your experience is like mine, nearly all. Bill Nichols has observed that the "hidden agenda of mastery and control" shaping Star Wars and military simulations is also evident in "the masculinist bias at work in video games." Both manifest the "masculine need for autonomy and control as it corresponds to the logic of a capitalist marketplace."[28] In the struggle between control and collectivity, virtual reality is contested ground. The two major fronts for research and development are military/government agencies on the one hand, and small entrepreneurial companies such as Autodesk and VPL on the other. While the U.S. Air Force uses flight simulators and virtual reality technology to prepare pilots for an invasion of Iraq, Jaron Lanier talks about the collaborative space created when multiple players interact to create a virtual world to which everyone contributes but that no one can dominate.[29]

The ethical orientations that Carol Gilligan identifies with male and female enculturations operate in virtual spaces no less than on playgrounds and in corporate offices.[30] The deep structures of virtual worlds are programmed in machine language and operate according to binary logic gates that follow linear decision paths. Layered over this deep structure is the matrix of possibilities of which the player is aware. What body form do you choose? How do you want the world to look? How do you want to interact with other players? In its collaborative aspects, virtual reality emphasizes connectivity, sensitivity to others' choices, open-ended creativity, free-wheeling exploration. It can, of course, be co-opted into masculinist ethics of competition and aggression. Even when this is not the case, the von Neumann architecture of the machine provides an underlying context of rule-governed choices that constitutes a masculinist subtext for the virtual world. It is not surprising, then, that writers who have extrapolated fictional worlds from virtual technology see them governed by masculinist ethics. Control is the dominant chord, subversion a minor but crucial intervention.

In *Gravity's Rainbow*, Thomas Pynchon wrote about the Zone, the freewheeling geopolitical space that opened for a brief time in Europe following the collapse of the Axis powers after World War II. In the Zone anything could happen, for power structures had not yet solidified their positions and ideologies were up for grabs. Virtual reality is a Body Zone, constructed not only through economic and geopolitical spaces but also through perceptual processing and neurological networks. Writers such as Vinge and Gibson, who are well aware of the technology's military potential, also see it as a space for political and cultural resistance. Both Case and Mr. Slippery, the protagonists of *Neuromancer* and "True Names," find themselves in opposition to the powers that be. For Case it is the Turing Police, who suspect he may be helping an artificial intelligence slip the shackles that keep it in check; for Mr. Slippery, the government functionaries who blackmail him into helping them fight an illicit user who is draining resources from the country's computer networks. The effect is a curious combination of totalizing power and exhilarating openness, as the names hint—Case recalling Wittgenstein's (and Pynchon's) "all that is the case," Mr. Slippery nominating the possibilities opened by slipping through the networks.

To understand the historical construction of the Body Zone, it is helpful to remember the predictions of Paul Virilio. Tracing the trajectory of speed through the twentieth century, Virilio foresaw that space would collapse into time, for when instantaneous communication and supersonic travel are commonplace across the globe, all cities exist in the same place—in time.[31] Tokyo is six hours away, New York three, Paris five. Accompanying this collapse was the coupling of the strategic capabilities of superpower military establishments and their consolidation with multinational corporations. Geopolitical boundaries take on different meanings when all territories lay open to instantaneous annihilation. Exocolonization, the deployment of military forces and economic imperialism against entities outside a country's borders, gives way to endocolonization, the appropriation of a country's own resources and population by the military-industrial complex. Latin American death squads are not anomalies, Virilio argues, but harbingers of the supplantation of exo- by endocolonization throughout the world. Thomas Pynchon corroborates Virilio's analysis in *Vineland*, where the narrator repeatedly observes that the populace of North California is subjected to drug raids and secret incarcerations *as of it were a Third World country.*[32]

The sense that the war has been carried to the home front is intensified by the suspicion that more than border patrols are involved. Also implicated is the blood-brain barrier. Endocolonization takes place not only through surveillance and terrorizing of the native population, but also through the ''colonization'' of ''wild-type'' genes (these are technical terms in genetic engineering) by retroviruses that supplant and usurp the native material.[33] The implosion of body politics into the interior of the body is given forceful expression by Greg Bear in *Blood Music*, where a nerdish engineer combines cybernetics and genetic engineering to invent ''bio-logic'' cells, microorganisms capable of intelligent decision making.[34] Going through several generations in a matter of hours, they evolve with exponential speed. By the time they escape the laboratory, they are already highly organized and intelligent. Within days they have mutated sufficiently to be able to decompose their host organisms, and humans everywhere disintegrate into cell colonies. As the biologic cells begin retrofitting the planet for their use, human beings become as rare as aardvarks, preserved by the cells as an endangered species. Nearly half a century ago, Norbert Wiener intuited that a possible implication of the shift to a cybernetic paradigm was the redefinition of the operative unit for survival and cooperation from macroorganisms to the microorganisms of which they are composed. *Blood Music* takes that intuition to its logical end.

The world has outrun Virilio's predictions. The contraction of external space did not in fact signal the end of spatiality, but rather its reconstitution on the other side of the computer screen and in the dark interior of the body. The new techniques of scientific visualization extend into the endospaces of the body as well as the cyberspaces of virtual reality.[35] The two are connected by more than the technology that unites internal perception to external computer. They are also articulated together through their social construction as areas newly available for colonization. In the scramble for power and control over these rich territories, there is still a place for wildcat entrepreneurs who buck the system with very little more than the quick reflexes that are, paradoxically, also part of the territory up for grabs. Thus in *Neuromancer*, when Case is caught stealing information from an employer, he is chemically altered so that he cannot enter cyberspace;

when another employer wants his services, the first step is to reconstruct his nervous system chemically and surgically, albeit with a built-in time bomb to ensure his loyalty to the project. Body politics is played within, as well as through, the bodies that engage in politics.

The Mirror of the Cyborg

The play between surface and depth as the computer screen opens into the high-dimensional projections within is worth dwelling on. Scott Bukatman has written about "terminal identity" as an "unmistakably doubled articulation in which we find both the end of the subject and a new subjectivity constructed at the computer station or television screen." He links the development of terminal identity to the "invasion and mutation of the body, the loss of the control, and the transformation of the self into Other," finding these mutations in such characteristically cybernetic works as William Burroughs's fiction and David Cronenberg's films (*The Fly* and *Videodrome*).[36] The simultaneous estrangement of the self from itself and its reconstitution as Other suggests that the diffusion of subjectivity through the cybernetic circuit constitutes a second mirror stage, the Mirror of the Cyborg.

As Lacan theorized it, the first mirror stage marks the initiation of the subject into language, the realm of the symbolic, and into the deferral and continuing lack that constitutes the play of signifiers.[37] The dialectic between absence and presence is central to Lacan's theory, as it is to much of deconstruction. The second mirror stage assumes that the speciousness of presence has been demonstrated and moves beyond it. Its central dialectic is between randomness and pattern. Constructing the subject through the flow of information that circulates within and around the system, it marks objects through patterns of assembly and disassembly rather than through the physical boundaries that are specularly recognized in the Lacanian mirror. Language gives way to the more general concept of messages-in-the-circuit. Communication takes place not only through words and syntax but also through the manipulation of cyberspace parameters. In cyberspace you do not necessarily need to *describe* how you see the world; you can visually and kinesthetically create it.[38]

In the Mirror of the Cyborg, anxiety about identity centers not on lack but on informational patterns that must cohere for continuity of the subject to be assured. The disaster corresponding to castration is flatlining, the dispersal of the pattern that represents the self. These speculations suggest that it is possible to rewrite Lacanian psycholinguistics as cyberlinguistics. The reinscriptions are summarized as follows:

Psycholinguistics	*Cyberlinguistics*
absence/presence	randomness/pattern

arbitrary relation of signifier to signified	arbitrary relation of message element to code
play of signifiers	random access memory
sliding/floating signifier	virtual memory
lack	noise
phallus	electroencephalogram (EEG)
castration	flatlining
repetition	redundancy
imaginary	physical
symbolic	virtual

In the construction of terminal identity, the play between two- and three-dimensional figures is extensive and complex. Highly charged sexual signifiers unfold differently in three-dimensional spaces for male and female. Extrusions and cavities take on gender identifications that create complex symbolic structures involving more than the phallus, as Irigaray and Cixous have insisted in their rewritings of Lacan.[39] In the cyborg mirror, three-dimensionality is reconstituted only after the encounter with the two-dimensional surface of the screen, which preexists before the virtual world opens and lingers after it has faded. Flatlining is a two-dimensional phenomenon, marking the screen as the juncture between the body, vulnerable to attack and decimation through physical means, and the cyborg puppet, vulnerable to destruction through the informational pattern that constitutes it. As gendered patterns of concavity and convexity move through the surface of the screen, they become more arbitrary, subject to rearrangements and reassemblies that are bound by informational rather than physical constraints. Thus the fictional worlds of cyberspace are replete with androgynous figures, from the warrior heroine of *Neuromancer* to the woman pirate of Kathy Acker's *Empire of the Senseless*.[40]

The additional dimensions that open beyond the specular reflections of the screen, reinforced by the fuller range of sensory feedback, give the Mirror of the Cyborg different dynamics from the Lacanian mirror. Moving into cyberspace binds subject and object positions together in a reflexive dynamic that makes their identification problematic. The putative subject is the consciousness embodied in a physical form, while the object is the puppet behind the screen. Since the flow of sensory information goes in both directions, however, the puppet can also be seen as the originary point for sensations. Along with many others who have experienced this technology, I found this ambiguity one of cyberspace's most disturbing and arresting features. Cyberspace represents a powerful challenge to the customary construction of the body's boundaries, opening them to transformative configurations that always bear the trace of the Other. The resulting disorientation can function as a wedge to destabilize presuppositions about self and Other.

In their negative manifestations, the self's boundaries act as symbolic structures that attack and denigrate whatever is outside and therefore different from the self, as if they were immune systems projected outside the skin and left to run amok in the world. When these dynamics prevail, the Other is either assimilated into the self to become an inferior version of the Same or remains outside as a threatening and incomprehensible alterity. So women are constructed as castrated men or Medusa figures; blacks as inferior whites or cannibalistic devils; the poor as lazy indigents or feral criminals. Conflating self and Other, the Mirror of the Cyborg brings these constructions into question. The metaphor of colonization should be taken seriously, for it suggests how we can use cyberspace to consolidate and extend lessons learned from postcolonialism. One can imagine scenarios in which the Other is accepted as both different *and* enriching, valued precisely because it represents what cannot be controlled and predicted. The puppet then stands for the release of spontaneity and alterity within the feedback loops that connect the subject with the world, as well as with those aspects of sentience that the self cannot recognize as originating from within itself. At this point the puppet has the potential to become more than a puppet, representing instead a zone of interaction that opens the subject to the exhilarating realization of Otherness valued as such.

Applied to the physical world, this realization values it for its differences from the virtual world—its incredibly fine structure, sensory richness, material stability, and spontaneous evolution. The positive seduction of cyberspace leads us to an appreciation of the larger ecosystems of which we are a part, connected through feedback loops that entangle our destinies with their fates. Bill Nichols says it best: "The cybernetic metaphor contains the germ of an enhanced future inside a prevailing model that substitutes part for whole, simulation for real, cyborg for human, conscious purpose for the decentred goal-seeking. . . . The task is not to overthrow the prevailing cybernetic model but to transgress its predefined interdictions and limits, using the dynamite of the apperceptive powers it has itself brought into being."[41] Apparently writing with no knowledge of cyberspace, Nichols nevertheless clearly sees the power of cybernetics as a metaphor. With cyberspace it becomes a representational space as well, simultaneously both model and metaphor. Hailing us on multiple levels, connecting physicality with virtuality, it opens new vistas for exploration even as it invites us to remember what cannot be replaced.

Notes

1. A contrary view is strongly argued by Roger Penrose, *The Emperor's New Mind: Concerning Computers, Minds, and the Laws of Physics* (New York: Oxford University Press, 1989, especially 373-447). O. B. Hardison, Jr., proclaims the end of the body in *Disappearing Through the Skylight: Culture and Technology in the Twentieth Century* (New York: Viking, 1989). His rhetoric typifies the postmodern fantasy of leaving the body behind.

2. Ed Regis, *Great Mambo Chicken and the Transhuman Condition: Science Slightly over the Edge* (Reading, Mass.: Addison-Wesley, 1990), 7.

3. Michel Serres, *Hermes: Literature, Science, Philosophy*, ed. Josué V. Harari and David F. Bell (Baltimore: Johns Hopkins University Press, 1982), 71-83.

4. Jean Baudrillard, *Simulations*, trans. Paul Foss, Paul Patton, and Philip Beitchman (New York: Semiotext[e], 1983), 1-78.

5. Fredric Jameson, *The Political Unconscious* (Ithaca, N.Y.: Cornell University Press, 1981).

6. Donna Haraway, ''A Manifesto for Cyborgs: Science, Technology, and Socialist Feminism in the 1980's,'' *Socialist Review* 80 (1985): 101.

7. Bill Nichols, ''The Work of Culture in the Age of Cybernetic Systems,'' *Screen* 29 (Winter 1988): 44.

8. The most complete history to date is found in Howard Rheingold, *Virtual Reality: The Revolutionary Technology of Computer-Generated Artificial Worlds and How It Promises and Threatens to Transform Business and Society* (New York: Simon and Schuster, 1991). See also Myron W. Krueger, ''Artificial Reality: Past and Future,'' *Multimedia Review* 1 (Summer 1990); and *Artificial Reality* (Reading, Mass.: Addison-Wesley, 1983), 1-28.

9. William Gibson, *Neuromancer* (New York: Ace, 1984).

10. John Walker, ''Through the Looking Glass: Beyond 'User' Interfaces,'' *CADalyst* (December 1989): 42.

11. Randall Walser, ''On the Road to Cyberia: A Few Thoughts on Autodesk's Initiative.'' *CADalyst* (December 1989): 43.

12. Michael Spring, ''Informating with Virtual Reality,'' *Multimedia Review* 1 (Summer 1990): 10-12.

13. Ibid., 11.

14. Scott Fisher, ''Personal Simulations and Telepresence,'' *Multimedia Review* 1 (Summer 1990): 24.

15. In Colin Greenland, ''A Nod to the Apocalypse: An Interview with William Gibson,'' *Foundation* 36 (Summer 1986): 5-9.

16. Joseph Henderson, ''Designing Realities: Interactive Media, Virtual Realities and Cyberspace,'' *Multimedia Review* 1 (Summer 1990), 50.

17. Ibid., 50-51.

18. Ibid., 51.

19. Walser, ''On the Road to Cyberia,'' 43.

20. John Perry Barlow, ''Being in Nothingness,'' *Mondo 2000* (Summer 1990), 42.

21. Ibid.

22. Clarke has developed the theme of eroticism in metamorphosis in ''Circe's Metamorphosis: Late Classical and Early Modern Neoplatonic Readings of the *Odyssey* and Ovid's *Metamorphoses*.'' *University of Hartford Studies in Literature* 21, no. 2 (1989): 3-20.

23. ''True Names'' is reprinted in Vernon Vinge, *True Names and Other Dangers* (New York: Baen, 1987).

24. Ibid., 47-48.

25. Howard Rheingold, ''Teledildonics: Reach Out and Touch Someone,'' *Mondo 2000* (Summer 1990), 52.

26. Vivian Sobchack has written eloquently about the necessity to remember that we are ''enworlded'' subjects in ''Postfuturism,'' in *Screening Space: The American Science Fiction Film*, 2d ed. (New York: Ungar, 1988), 223-306; and ''The Scene of the Screen: Toward a Phenomenology of Cinematic and Electronic 'Presence,' '' in *Materialität der Kommunikation*, ed. Hans U. Gumbrecht and K. Ludwig Pfeiffer (Frankfurt am Main: Suhrkamp, 1988). Working from a Heideggerian frame of reference, she constructs in the latter a phenomenology that traces a trajectory from photographic

nostalgia to a cinematic "thickening" of the present to an electronic flattening of temporality into an instant. Much of what she has to say about electronic culture is relevant to the present argument.

27. Tom DeHaven, *Freaks Amour* (New York: Penguin, 1986).

28. Nichols, "The Work of Culture," 43.

29. Lanier discusses this aspect of virtual reality in Kevin Kelly, "Virtual Reality: An Interview with Jaron Lanier," *Whole Earth Review* 64 (Fall 1989).

30. See Carol Gilligan, *In a Different Voice: Psychological Theory and Women's Development* (Cambridge, Mass.: Harvard University Press, 1982).

31. See Paul Virilio and Sylvère Lotringer, *Pure War*, trans. Mark Polizzotti (New York: Semiotext[e], 1983), 60.

32. Anthony Wilden makes this same point when he suggests that the real conflicts are not between one country and another but between the military-industrial complexes of all countries and ordinary people; see *The Rules Are No Game: The Strategy of Communication* (London: Routledge and Kegan Paul, 1987).

33. For a further explanation of genetic engineering techniques and the colonization metaphor, see N. Katherine Hayles, "Postmodern Parataxis: Embodied Texts, Weightless Information," *American Literary History* 2, no. 3 (1990): 394-421.

34. Greg Bear, *Blood Music* (New York: Ace, 1986).

35. Techniques of scientific visualization go beyond cyberspace, although virtual reality is part of the computer revolution in visualization. For a complete account, see Richard M. Friedhoff and William Benzon, *Visualization: The Second Computer Revolution* (New York: Harry N. Abrams, 1989).

36. Scott Bukatman, "Who Programs You? The Science Fiction of the Spectacle," in *Alien Zone: Cultural Theory and Contemporary Science Fiction Cinema*, ed. Annette Kuhn (London: Verso, 1990), 201.

37. Jacques Lacan, *Ecrits: A Selection*, trans. Alan Sheridan (New York: Norton, 1977).

38. Jaron Lanier has gone so far as to suggest that the kinesthetic manipulation of cyberspace will supplant language, making it an unnecessary and superfluous adjunct to virtual reality. This position ignores the underlying assembly language that governs the syntax of the computer program. It also fails to take into account that our sensibilities are formed through language, so that in this sense language pervades even nonlinguistic domains.

39. Luce Irigaray, *This Sex Which Is Not One*, trans. Catherine Porter and Carolyn Burke (Ithaca, N.Y.: Cornell University Press, 1985); and *Speculum of the Other Woman*, trans. Gillian C. Gill (Ithaca, N.Y.: Cornell University Press, 1985); Hélène Cixous and Catherine Clément, *The Newly Born Woman*, trans. Betsy Wing (Minneapolis: University of Minnesota Press, 1986).

40. Kathy Acker, *Empire of the Senseless* (New York: Grove, 1988).

41. Nichols, "The Work of Culture," 46.

Chapter 11

The Leap and the Lapse: Hacking a Private Site in Cyberspace

Alberto Moreiras

Thinking Cyberexcess

Octavio Paz remarked in 1967 that cybernetics came close to poetry in its use of universal analogy.[1] Virtual reality, grounded in the production of analogues aiming at the total illusion of reality, is an apotheosis of what the old metaphysicians called *analogia entis*. But a total illusion, insofar as it approaches completion in the realization of its essence, equivocates the real while at the same time breaking the ground of analogy.

Analogy must be founded. *Esse* founds the possibility of the universal analogy of the *entes*. Virtual reality, as the possibility of total replication, including the replication of the ground of analogy, forces the question: Is analogy analogical? Virtual reality, which I shall define as analogy of analogy, opens the abyss of ontotheology by radically soliciting the essence of ground. In that sense, virtual reality, as the future of technology, holds within itself the possibility of ungrounding technology. Virtual reality threatens the stability of the highest principle of technological being, the principle of sufficient reason, according to which there is nothing without a reason, there is nothing without a ground.[2]

A question that seemed settled at the height of the Cold War, namely, that our times were historically marked as the nuclear age, has now become undecided. Whether or not we think that the possibility of a nuclear confrontation has temporarily receded, the indecisiveness concerning the mark of the times has increased with the fall of the Berlin Wall; so has the claim of cybernetics and its password, *information*. In their realm of possibility, both cybernetics and atomic technology depend upon representational-calculative thinking, that is, the thinking that gives itself over to "the demand to render sufficient reasons for all representations."[3] In virtual reality, the principle of sufficient reason holds at its most extreme. Virtual reality is also the site of the most extreme withdrawal of what the principle of sufficient reason cannot comprehend.

If poetic experience, as Paz and also Jorge Luis Borges claimed at a certain moment, is an experience of analogical transcendence, then poetic thinking may no longer be sufficient to distinguish human thought from computer information-processing capabilities.[4] In a crucial sense, in and through the development of virtual reality, the poetic principle of tropological production is being absorbed today by cybertech. Does cybertech merely put tropology at the service of onto-theological (technical) reproduction? In other words, is cybertech contained within the reproductive mode proper to metaphysics, understood as ontotheology? Or does it hold another possibility?

If virtual reality is to be defined as an analogical transposition of the real, a trans(in)formation of the real working through analogy, then virtual reality is a metaphoric mode. But metaphor, depending as it does upon the division between the sensible and the nonsensible, "exists only within metaphysics."[5] However, there may be ways of dwelling within virtual reality that are nonmetaphoric, insofar as they come close to the end of metaphor.

If, like poetry, cybernetics can incorporate the real in its most extreme moments, as shining, objectified presence on the one hand, and as total withdrawal on the other, then cybernetics can also be interrogated analogically. By an *analogical interrogation* I mean a mode of questioning concerned with finding the point of articulation of presence and withdrawal in the technical system of representation. Can cybertech reflect on cybertech? I will attempt to think cyberspace as poetic space, and poetic space from the perspective of cyberspace. It remains to be seen whether or not analogy is the last principle of poetry and/or of the cybernetic real—that is, of virtual reality.

That cybernetics is complicitous with ontotheology remains undecided. On the sinister side, we read the dystopic projections of William Gibson and Bruce Sterling, who, in their novel *The Difference Engine*, imagine a so-called Modus Program, whose virtue would be to do away with the limitations embedded in the Leibnizian dream of finding a *characteristica universalis* in logical closure. The Modus Program, incorporating transfinite principles, will "form the bedrock of a genuinely transcendent meta-system of calculatory mathematics."[6] As a result, it will give the cyberengine a self-referential capacity. As the machine grows sufficiently large, what had up to then been a vicarious eye will develop an I: "The Eye at last must see itself."[7] An ultimate panopticon will be set in place. Onto-theology will have come to its radical completion through a most extreme form of simulation: the reality engine, the matrix of all human engineering, will take its long-announced position as First Subject. An apotheosis, completion will come as the exact reverse of the nuclear Armageddon: it will not be, at least not preeminently, a destruction, but a totally in-formed construction.

Other accounts, such as Donna Haraway's "A Manifesto for Cyborgs," substitute a euphoric, highly celebratory mood for the dejected and destitute one:

> From one perspective, a cyborg world is about the final imposition of a grid of control on the planet, about the final abstraction embodied in a Star Wars apocalypse waged in the name of defense, about the final appropriation of women's bodies in a masculinist orgy of war. From another perspective, a cyborg world might be about lived social and bodily realities in which people are not afraid of their joint kinship with animals and machines, not afraid of permanently partial identities and contradictory standpoints.[8]

For Haraway, high-tech culture offers the possibility of challenging phallogocentrism, but only if high-tech culture is accompanied by a refusal of victimization stories, all of which, whether explicitly or not, advocate "an anti-science metaphysics, a demonology of technology":[9]

> Every story that begins with original innocence and privileges the return to wholeness imagines the drama of life to be individuation, separation, the birth of the self, the tragedy of autonomy, the fall into writing, alienation; that is, war, tempered by imaginary respite in the bosom of the Other. These plots are ruled by a reproductive politics—rebirth without flaw, perfection, abstraction. In this plot women are imagined either better or worse off, but all agree they have less selfhood, weaker individuation, more fusion to the oral, to Mother, less at stake in masculine autonomy. But there is another route to having less at stake in masculine autonomy, a route that does not pass through Woman, Primitive, Zero, the Mirror Stage and its imaginary. It passes through women, and other present-tense, illegitimate cyborgs, not of Woman born, who refuse the ideological resources of victimization so as to have a real life.[10]

Haraway refuses resentment, and her position is active rather than reactive. Her politics of real life "insist[s] on noise and advocate[s] pollution, rejoicing in the illegitimate fusions of animal and machine."[11] Haraway places her emphasis on "disturbingly and pleasurably tight coupling," a coupling that would be far from traditional coition, pointing as it does against a metaphysics of the reproductive copula.[12] However, Haraway's manifesto for a radically nonessentialist, postgender world in cyberspace seems oblivious of its consequences. Antiessentialism has a short memory. In a sense, Haraway's celebration of the cyborg's subversion of identity within contemporary technology disregards the "withinness," the essential mark that the frame inscribes upon any enframed antiessence. Supposing that this disregard is not a consequence of nonknowledge, but rather an active blindness, an active oblivion, will it achieve what it is meant to achieve?

Cybertech, as the future of technology, is within the purview of the calculative-representational enframing of the world, and, as such, it is essentially to be understood within the scope of the principle of sufficient reason. In its short form the principle says: *nihil est sine ratione*, nothing is without reason. The apotheosis of analogical reason in virtual reality is such that, in virtual reality, everything is in virtue of *ratio* understood as proportionality. Analogical reason is the ground of virtual reality. From the perspective of virtual reality, nothing is without an analogue. Virtual reality renders the real as the mere possibility of replication, only awaiting the moment in which replication can double itself in self-replication. There is danger in this, as Gibson and Sterling see it, because the disappearance of the real can mean that the real has been sequestered. But there is also seduction, as Haraway sees it, because, in a world with no original, there is but the rhetorical effectiveness of translation. Is it possible to think beyond danger and seduction, or, even better, affirm both the seduction of danger and the danger of seduction?

Virtual reality challenges the human capacity to realize understanding of being. In virtual reality, artificial intelligence, familiar in its technical conspicuousness, reverts into the most unfamiliar obdurateness as it purports to replicate the human world, returning to us in the process worldliness as the most obstinate form of familiarity. Within virtual reality, there is no always-already, except in the merely privative mode; that is, virtual reality, even in the extreme form of total success at representation, cannot but perpetually enact the world as lost object. Within virtual reality, the worldliness of the world unconceals itself, even if in the form of absence. To ask whether nonrepresentational thinking can help us deal with the phenomenon of virtual reality is also to ask whether or not virtual reality can offer an opening onto critical-historical thinking. It is not only to ask whether virtual reality can be experienced as a possibility for a thinking of the Outside, but also whether it affords the possibility of a break. It would have to be a break away from the calculative-representational frame that originated it. It would also be a break into a region of thinking where the calculative-representational frame would not be merely ignored or forgotten, but brought to account for itself.

Can we define a task of thinking that would refuse to believe itself above and beyond technique? This question, which has plagued contemporary philosophy, is also to be found within poetic thought.[13] It recurs in several stories written by Borges in the 1940s, and particularly in "El Aleph," which presents one of the earliest literary treatments of the kind of technological space that we now call cyberspace.

The space defined by the object called Aleph is not properly speaking cyberspace, understood as the locus where the human interfaces with artificial intelligence machines. Nevertheless, in Borges's text the Aleph is announced analogically as the site of encounter where "modern man" meets robotic control of reality.[14] If cybernetics comes from the Greek word *kybernetes*, meaning pilot or governor of a ship, and if it designates the steering function of the brain-within-machines, then the antagonist in Borges's story talks about the cybernetic man when he observes that, for the moderns, "the act of travel [is] useless." The old pilot of the ship can now reach the world from his own study, using "telephones, telegraphs, phonographs, radiotelephone apparatus, cinematographic equipment, magic lanterns."[15] Action at a distance, telepraxis, would create the space of the cybernetic human, cyberspace. As a transposition of this cyberspace, analogically, the text gives us the uncanny apparatus properly called Aleph.

An Aleph is "one of the points in space containing all points."[16] It can be directly experienced, but it cannot be translated; it can be indicated, but it cannot be expressed. It is a radical place of disjunction, where language breaks down. Borges calls it "the ineffable center of my story," where there occurs his "despair as a writer."[17] As it can be named only analogically, it thereby grounds the insufficiency of analogy. It is the site of the real, where the real announces itself in withdrawal. It is a *punctum*, in the Latin sense that Roland Barthes emphasizes: a place where the trace of presence is poignantly felt in default, a site of mourning, a private site.[18]

As the narrator is lying down, alone, in the basement of his late beloved's house, uncannily undergoing an experience of encryptment within the analogue of Beatriz's dead body (the house is about to be demolished), he sees the Aleph. I will quote only the end of his description:

> I saw tigers, emboli, bison, ground swells, and armies; I saw all the ants on earth; I saw a Persian astrolabe; in a desk drawer I saw (the writing made me tremble) obscene, incredible, precise letters, which Beatriz had written Carlos Argentino; I saw an adored monument in La Chacarita cemetery; I saw the atrocious relic of what deliciously had been Beatriz Viterbo; I saw the circulation of my obscure blood; I saw the gearing of love and the modifications of death; I saw the Aleph from all points; I saw the earth in the Aleph and in the Aleph the earth once more and the earth in the Aleph; I saw my face and my viscera; I saw your face and felt vertigo and cried because my eyes had seen that conjectural and secret object whose name men usurp but which no man has gazed on: the inconceivable universe. I felt infinite veneration, infinite compassion.[19]

"Desde todos los puntos vi en el Aleph la tierra y en la tierra otra vez el Aleph y en el Aleph la tierra": in this frenzied, chiastic doubling of analogy, this analogy of analogy, or abysmal experience wherein the point that contains every point must perforce contain itself and therefore also reveal itself as the uncontainable, the ground of analogy breaks in excess. The excess exceeds analogy. Borges mentions ''inconceivable analogies'' in trying to equate the Aleph with the mystical experience of divinity, which Alanus de Insulis had described by calling it ''a sphere whose center is everywhere and whose circumference is nowhere.''[20] The inconceivability of analogy is here the mark of an excess with respect to analogy. This excess connotes an experience of the real-in-withdrawal that can perhaps be located in what I will call ''a private site.''

As *privare* is in Latin to deprive, to take away, to set apart, it also consequently means to release from common use and therefore to secure into its own. A private site is a site in need, where what lacks is at the same time protected. As set apart, it stands on its own. On its own, it lacks that from which it has been secured. It is a site of releasement where excess can be rendered as recess. In recess, in withdrawal, the private stands secluded, out of reach. Concealed, always concealing, it is experienced as a site of loss.

The mystical experience turns toward divinity, but the poetic holds fast to the necessity of expression, in which recess, as withdrawal, as the end of analogy, as the abyss of tropology, remains a vanishing point and not a point of advent. Because the point vanishes, Borges is led to conclude: ''The Aleph in the Calle Garay was a false Aleph.''[21]

At the end of analogy, when language opens toward the real as withdrawal, poetic thinking thinks the nothing as withdrawing excess. If the nothing as withdrawing excess is revealed in writing, it is revealed as a break in tropology. But tropology names literary technique. Now, we have to ask, will it work in cybertech? What experience of thinking does cybertech make possible, even necessary?

The Want of the Letter

Hacking, the word commonly used to describe the acts of those who manage to clear their way into locked computer systems, originally carried the meaning of severing with repeated blows, clearing by cutting away vegetation. A computer hacker makes a clearing for her- or himself. The addictive quality of hacking could be emblematized in the words of Dirk-Otto Brzezinski, one of the hackers implicated in the Project Equalizer espionage case, who told his judge: ''I was never interested in the contents. Just in the computers themselves.''[22] His remark does not replicate the common rhetorical distinction between form and content within a literary text; rather, it points to a different realm of experience. The distinction between ''contents,'' the actual information stored within a given computer system, and ''computers themselves,'' referring to something more than a mere machine, raises the question of excess anew.

The hacker wants to break in. Breaking in is the addictive principle of hacking, so that the clearing made possible by hacking can manifest itself. The ''computers themselves'' are the engines that make breaking in possible. More radically, the computers themselves are the clearing. The computer-as-clearing opens onto cyberspace as transgressive space, the space beyond the break. Howard Rheingold, in *Virtual Reality*, comments, ''It is a place, all right. What kind of place it is, is a big question''; he goes on to quote Gibson's definition of cyberspace from the 1984 novel *Neuromancer*:

> Cyberspace. A consensual hallucination experienced daily by billions of legitimate operators, in every nation, by children being taught mathematical concepts. . . . A graphic representation of data abstracted from the banks of every computer in the human system. Unthinkable complexity. Lines of light ranged in the nonspace of the mind, clusters and constellations of data. Like city lights, receding.[23]

Cyberspace is a receding space, a withdrawing space, a space as recess. To break into the perpetual recession: such is the addiction that dreams cyberspace as a private clearing for its human interfacers. It produces anxiety, as it is a melancholic exercise in endless loss.[24]

At the end of his book, Rheingold devotes a few pages to speculation on popular cyberdreams such as ''teledildonics'' (sex at a distance) and ''electronic LSD.'' Rheingold makes it clear that, although both technologies remain undeveloped, they are not beyond the pale of technical prediction. One example:

> If you can map your hands to your puppet's legs, and let your fingers do the walking through cyberspace, as it is possible to do in a crude way with today's technology, there is no reason to believe you won't be able to map your genital effectors to your manual sensors, and have direct genital contact by shaking hands. What will happen to social touching when nobody knows where anybody else's erogenous zones are located?[25]

I can't wait. But the sheer possibility of perpetual overdose has on its flip side the poisonous presence of deprivation. Rheingold says that ''privacy and identity and intimacy will become tightly coupled into something we don't have a name for yet.''[26] Or rather: the name is, will be, unavowable.

Cyberexcess—as writing once did—will kill the need for memory. Excess as primary manifestation links cyberspace and the space of writing. In ''El Zahir,'' another story from the 1949 collection, Borges retells the myth of Fafnir and the treasure of the Nibelungs.[27] If Fafnir's mission is to keep watch and therefore to guard the existence of the treasure, that treasure can be accessed only by killing Fafnir. And what kills Fafnir, the sword Gram, bears the name of writing, or of the letter. Gram opens the treasure, gives the treasure, but at the same time Gram kills what secured the treasure. The letter releases what it was supposed to secure, the gift of memory. The letter, as excess, is also a form of want.

Clearing into cyberspace radically engages cyberware as a writing machine. Cyberspace is not a letter, but our relationship to it has the structure of our relationship to the letter in the following sense: primarily understood as an entrance into analogical production, clearing into cyberspace is also at the same time an excessive activity that takes analogy to a breaking point. In the break, cyberspace is felt as a wanting space, a space of default. Cyberspace is a site of disjunction, where analogical production comes to find the limits of analogy. The experience of the limit that cyberspace affords is an anxious, addictive experience in which the real appears as withdrawal and loss. Cyberexperience is in that sense akin to the experience that Borges tells of in ''El Aleph.''

The want of the letter is ultimately the theme of ''El Aleph.'' An Aleph is ''the first letter of the alphabet of the sacred language,'' and as such a symbol of ''pure and unlimited divinity.''[28] That it lacks even as it gives itself, that it gives itself in lack, that is what the principle of reason cannot account for. Through reading ''El Aleph'''s relationship to woman, in the following section of this essay I will try to show that, at a certain point, the poetic need for ontotheological reproduction breaks down. Such a break is a function of writing itself as ''technique.'' A certain analogy between writing and cybertech obtains even as both announce the end of analogy. This end of analogy, far from being a point of ultimate disjunction between philosophic, poetic, and technical thinking, is a gathering point, where the task of thinking can retrieve the possibility of going beyond the private.

The Lapsarian Experience

In ''Two Words for Joyce,'' Jacques Derrida talks about ''two manners, or rather two greatnesses, in this madness of writing.'' One of them, for which apparently no instance is given, is the writing of the gift: ''There is first of all the greatness of s/he who writes in order to give, in giving, and therefore in order to give to forget the gift and the given, what is given and the act of giving, which is the only way of giving, the only possible—and impossible—way.''[29] The second greatness is that of a ''hypermnesiac machine'' such as the Joycean text (or the textuality given in Borges's Aleph, or, even more pointedly, the cybertext): ''You can say nothing that is not programmed on this 1000th generation computer—Ulysses, Finnegans Wake—beside which the current technology of our computers and our micro-computerified archives and our translating machines remain a bricolage of a prehistoric child's toys.''[30] If the first kind of writing places itself by definition in a paradoxical gratitude involving not only the writer and the reader but also the matter at hand, whatever that is, the second kind of writing involves not gratitude, but its opposite, ''resentment and jealousy.'' ''Can one pardon this hypermnesia which a priori indebts you, and in advance inscribes you in the book you are reading? One can pardon this Babelian act of war only if it happens already, from all time, with each event of writing, and if one knows it.''[31] Is cyberspace implied, from all time, in each event of writing?

If the hypermnesiac machine, the 1000th generation computer, acts with each event of writing, we may wonder whether the writing of the gift also operates every time. And what about their mutual coimplication, and the relation, in writing, between gratitude and resentment? Doesn't the impossible combination of those affects organize the melancholic state? In virtual reality, is there one ''greatness'' without the other? Is there a gift in cyberspace? Or is there only a negation of the gift? Are we but resentment freaks, who love the debt, and are grateful for what pains us? These questions also need to be asked of ''El Aleph,'' and of the kinds of writing it contains.

Its narrator takes a leap into the excessive region of total, hypermnesiac presence. Accounting for that experience organizes ''El Aleph'''s writing field. As the narrator cannot replicate the ''ineffable center'' of his experience, he must give himself over to a sort of lapsarian writing: a writing that can only refer to a fall into that which exceeds its possibilities of expression, a writing understood as the site of the fall into the withdrawing recess of expressibility.[32]

In ''El Aleph,'' writing indicates what has slipped away, that is, what has withdrawn and, in withdrawing, has made itself obtrusive, and has in such a way come into paradoxical presence. Writing, thus understood, does not essentially differ from the cybernetic experience of virtual reality. Cybertech, in its extreme form, opens the possibility of an experience of the ground of technology as withdrawing ground—that is, not the ontotheological ground that secures every object into the shelter of a foundation, but the receding ground that releases the real as vanishing materiality, beyond analogy, beyond memory.

Borges's writing is essentially metadiegetic, a telling of telling. For Borges, ''we can mention or allude, but we cannot express.''[33] For Borges, writing is never more, or less, than an indication. In ''El Aleph'' Borges compares critical writing to the activity of those persons ''who dispose of no precious metals, nor steam presses, nor rolling presses, nor sulphuric acids for minting treasures, but who can indicate to others the site of a treasure.''[34] ''El Aleph'' is precisely that kind of gesture: an indication of an ineffable center that cannot be named as such, but only analogically. Borges's description of the Aleph fails to give the Aleph: the Aleph cannot happen in writing, for writing is the place of its lapse. Writing organizes the want of the letter, and can give only what it does not have, like virtual reality, as in virtual reality the world can be experienced only as the lost object of analogy.[35]

Within the system of ''El Aleph,'' writing occurs on a dead woman's body. As Beatriz's house houses the Aleph, Beatriz's house is the site of the gift. However, as the Aleph can only be forgotten—all Alephs are false Alephs—Beatriz's house is also the site of resentment and jealousy. Writing copes with both gratitude and resentment on indicating the lost object: an object that can be mentioned or alluded to, but that cannot be expressed, for it remains in excess. Borges's writing is an attempt to seduce the excess into self-revealing, an anxious attempt to turn the lapse into a leap, to make withdrawal come, as such, into presence. At the same time, however, Borges marks another possibility of writing, whose parallel possibility we can also find in cyberspace.

At the very beginning of "El Aleph," the narrator tells us that his visits to Calle Garay on the day of Beatriz's birthday were a ceremony of mourning. By returning to Beatriz's house, the narrator gives himself over to mournful memory: "Now that she was dead, I could consecrate myself to her memory, without hope but also without humiliation."[36] Beatriz's death is therefore initially understood as affording a certain chance, involving a double renunciation—on the one hand, the renunciation of Beatriz as gift; on the other, the renunciation of the torturing possibilities of jealousy and resentment. That chance is the chance of memory, understood as consecration, that is, self-offering. The narrator wills such an offering to be free of poignancy, of pain. By keeping Beatriz in his memory the narrator will live in the memory of Beatriz: a self-willed self-giving, nothing else, studied, and contained.

Every time Borges's narrator arrives at the house in Calle Garay he is made to wait. There, "in the twilight of the overladen entrance hall," he

> would study, one more time, the particulars of [Beatriz's] numerous portraits: Beatriz Viterbo in profile, in color; Beatriz wearing a mask, during the carnival of 1921; Beatriz at her First Communion; Beatriz on the day of her wedding to Roberto Alessandri; Beatriz a little while after the divorce, at a dinner in the Club Hípico; Beatriz with Delia San Marco Porcel and Carlos Argentino; Beatriz with the Pekingese . . . ; Beatriz . . . smiling, her hand under her chin.[37]

At the threshold, before being summoned to the depths of the house in whose cellar he will find a very different rapport to the images of Beatriz, the narrator chooses, explicitly, a way of relating to those photographs consonant with his desire to live in memory of Beatriz "without hope but also without humiliation." The narrator's conscious investment in Beatriz's death is made according to an economy of limited expenditure: or rather, an economy of nonexpenditure, an aberrant economy of repression in which, however, mourning follows its normal process of completion. In this studious relationship to Beatriz we find one of the possibilities of experience that virtual reality may have to offer: a guarded experience in which everything is made to function by analogy, through calculative, mimetic memory. By apparently resisting jealousy and resentment, this mimetic memory essentially yields to jealousy and resentment, since it refuses to hold itself open to the anxious possibility of the gift.

I cannot go here into the aspects of the Borgesian text in which that studious relationship to the monument is linked to the practice of a certain kind of reproductive literature. The writings of Carlos Argentino Daneri (who acts, in spite of his name, as the narrator's Virgil) exemplify a mimetic literature of exhaustion, regulated by the will to express the expressible, to saturate the field of the real. Against them, Borges's metadiegesis opts for the breaking of mimesis: the (non) expression of the inexpressible, the fissure in consciousness. But both possibilities, the mimetic possibility of replication and the lapsarian possibility of release, are also the two sides of the cybernetic interface.

Daneri, the narrator's Virgil, takes him to the cellar of the house in Calle Garay, and makes him lie in a "dorsal decubitus" position: "Now, down with you. Very shortly you will be able to engage in a dialogue with *all* of the images of Beatriz."[38] The Aleph, as the point containing all points, will be given as the site for the essential breaking of the studious reproduction of the real. In the Aleph, the real returns as what is essentially out of reach, beyond appropriation. Beatriz, who shows up in the narrator's account as the receiver of obscene letters, and as the atrocious corpse within La Chacarita's funeral monument, returns blindingly as the occasion, the chance, for infinite jealousy and resentment, even as her house, her memory, is also the region of the endless lucid gift. With it, with them, the narrator lives in memory of Beatriz, in her memory as total memory, no longer guarded, no longer self-willed. He could repeat what Barthes said: "I could live without [her] (we all do it, sooner or later); but the life who for me remained would be, certainly and until the end, *unqualifiable* (without quality)."[39] When our narrator comes out of his experience he feels, curiously and almost impossibly, not only awe and pity, but also, for a moment, "indifference."[40]

After the protagonist in Borges's story has experienced the Aleph, after he has had his tragic immersion in infinite awe and pity, he comes out of it in deep shock, and refuses to share his experience: "I refused, with suave energy, to discuss the Aleph." He has, at that point, decided to take the gift, and he has used it to placate the envy he feels for his rival Carlos Argentino. The gift becomes obsessive: "I was afraid that I would never be quit of the impression that I had 'returned' [*Temí que no me abandonara jamás la impresión de volver*]."[41] But oblivion sets in. The narrator can then come to the conclusion that the Aleph was false. Since it was false, it goes back into concealment, into "the innermost recess of a stone."[42] The narrator can once again experience the world outside analogy. Oblivion, and not the Aleph, is ecstasy. Oblivion is the gift, as it is the (broken) end of mourning. "There is first of all the greatness of s/he who writes in order to give, in giving, and therefore in order to give to forget the gift and the given."[43] Oblivion has to be gained, and it is therefore an active oblivion, in the sense of an active opening toward the work of the gift.

We can argue whether this kind of writing is still subject to phallic bliss, or whether, by announcing the end of analogy, it has explicitly put an end to the ontotheological need for self-reproduction. Lapsarian writing does not want more of the same: rather, what it wants cannot be had. The leap, which is not the leap of the narrator as character, but that of the narrator as narrator, as metadiegetic writer, is taken not toward the treasure, but toward the site where the treasure vanishes, which is the private site. The site where the treasure vanishes is, however, the site of closest proximity to the treasure: the region of its recess, a region both dangerous and seductive, the region where the private opens itself to the unavowable.

The leap into the unavowable is also the most radical possibility of the cybernetic human. Within cyberspace, two experiences are given: the mimetic experience, which is the experience of cyberspace as a space of analogical production; and the lapsarian experience, which comes to the end of analogy. As in "The Aleph," those two experiences can also be explained by reference to woman.

The expression "cyborg envy" has been used to talk about the inversion of the classical "penis envy" taking place in the longing for cyberspace. Stone notes that the cybernetic mode "shares certain conceptual and affective characteristics with numerous fictional evocations of the inarticulate longing of the male for the female."[44] In "cyborg envy" we long to become woman. In the cybernetic act, "penetration translates into envelopment. In other words, to enter cyberspace is physically to *put on* cyberspace. To become the cyborg, to put on the seductive and dangerous cybernetic space like a garment, is to put on the *female*."[45]

To understand entering cyberspace as the act of putting on something or other, someone or other, is to understand cyberexperience as essentially mimetic in nature. But we have seen that entering the Aleph is not to become Beatriz. Entering the Aleph, and entering cyberspace, can be felt as experience of a break, and therefore experience of distance and of loss, having nothing to do with envelopment, since they occur in the real, like danger, and seduction. "Putting on" the female, as a mimetic experience in cyberspace, is on the side of the studious, guarded relationship to mourning that Borges's narrator experiences at the threshold of Beatriz's house.

In the Aleph experience woman figures as the ground of the gift, but also as the ground of the infinite withdrawal of the gift, which is the ground of memory and oblivion. In computer hacking the contents are much less interesting than the puncturing of the computers themselves, as ground of memory and as total resistance to memory. It may then be that entering cyberspace can offer the possibility of being poignantly enveloped by the self-revealing withdrawal of the real: an experience of the loss of otherness that does not result in a reappropriation of sameness, but in a disjunction that manages a particular form of juncture, letting juncture come into its own.

The lapsarian experience is the most radical experience of cyberspace. Antimimetic, because it comes to the end of mimesis, it may use the mimetic engine up to a certain point. If "putting on the female" means, for Stone, not just to replicate or subvert penis envy, but to engage in a strategy of replication the sense of which is to release lapsarian writing into its own, then it might also mean to go beyond the principle of reason, into an experience of the real that, having already given up the need for appropriation of the gift, is no longer naive enough to assume that the 1000th generation machine can really read us all. For even if it wants it, it cannot have it. This refusal is also an act of love, reasonable too, though melancholic.

The extent to which oblivion needs to have a reason is the extent to which the Aleph, and with it cyberspace, is always already implied in every act of writing and of reading. The lapse, without which there is no leap, is not a mere abyss, not just an inversion of the principle of reason. The reason, the ground, of oblivion, is also the ground of lapsarian writing. Oblivion forsakes analogy, and brings the end of representation within the possibility of an excessive/recessive call of thinking.

Virtual reality, as a mere replication of possibilities, readily affords to be used as a mimetic tool for analogical exhaustion. In virtual reality, we can put on woman, no less than we can put on anything we have or anything we do not have. In this mood, we are fully within the space of the calculative-representational frame expressed by the Leibnizian principle of sufficient reason. But cyberspace also opens itself to the lapsarian experience: at the end of analogy that (un) grounds all analogy, cyberspace shelters a gift for which we can never fully find, or render, a reason.

Notes

1. Octavio Paz, *El arco y la lira*, 3d ed. (Mexico City: Fondo de Cultura Económica, 1986), 33.

2. For a discussion of the importance of the Leibnizian principle for technological thinking, see Martin Heidegger, *The Principle of Reason*, trans. Reginald Lilly (Bloomington: Indiana University Press, 1991).

3. Ibid., 33.

4. See Jorge Luis Borges, ''El arte narrativo y la magia,'' in *Prosa completa*, 2 vols. (Barcelona: Bruguera, 1980), 1.163-70.

5. Heidegger, *The Principle of Reason*, 48.

6. William Gibson and Bruce Sterling, *The Difference Engine* (New York: Bantam Spectra, 1991), 421.

7. Ibid., 429.

8. Donna Haraway, ''A Manifesto for Cyborgs: Science, Technology, and Socialist Feminism in the 1980's,'' in *Feminism/Postmodernism*, ed. Linda J. Nicholson (New York: Routledge, 1990), 196.

9. Ibid., 223.

10. Ibid., 219.

11. Ibid., 218.

12. Ibid., 193.

13. A classically Heideggerian question, it is also, arguably, the question of deconstruction. See Jacques Derrida, *Mémoires: Pour Paul de Man* (Paris: Galilée, 1989), 109.

14. Jorge Luis Borges, "The Aleph," in *A Personal Anthology*, ed. and trans. Anthony Kerrigan (New York: Grove, 1967), 140; and *Prosa completa*, 2.113.

15. Borges, "The Aleph," 140; and *Prosa completa*, 2.114.

16. Borges, "The Aleph," 146; and *Prosa completa*, 2.119.

17. Borges, "The Aleph," 148; and *Prosa completa*, 2.121.

18. Roland Barthes, *La chambre claire: Note sur la photographie* (Paris: L'Etoile/Gallimard/Seuil, 1980), 48-49ff.

19. Borges, "The Aleph," 151; and *Prosa completa*, 2.122.

20. Borges, "The Aleph," 149; and *Prosa completa*, 2.121.

21. Borges, "The Aleph," 153; and *Prosa completa*, 2.124.

22. Quoted in Katie Hafner and John Markoff, *Cyberpunk: Outlaws and Hackers on the Computer Frontier* (New York: Simon and Schuster, 1991), 240.

23. In Howard Rheingold, *Virtual Reality: The Revolutionary Technology of Computer-Generated Artificial Worlds and How It Promises and Threatens to Transform Business and Society* (New York: Simon and Schuster, 1991), 16.

24. See James Joyce, *Finnegans Wake* (Harmondsworth: Penguin, 1986), 611-12, for an association of anxious melancholy and a vision in which "all objects allside showed themselves," and a lot more.

25. Rheingold, *Virtual Reality*, 352.

26. Ibid.

27. Borges, *Prosa completa*, 2.81.

28. Borges, "The Aleph," 153; and *Prosa completa*, 2.124.

29. Jacques Derrida, "Two Words for Joyce," in *Post-Structuralist Joyce: Essays from the French*, ed. Derek Attridge and Daniel Ferrer (Cambridge: Cambridge University Press, 1984), 146.

30. Ibid., 147.

31. Ibid.

32. For a notion of the "lapsus" as the organizing space of reading and writing, see Jacques Lacan, *Encore: Le séminaire XX*, ed. Jacques-Alain Miller (Paris: Seuil, 1975), 37.

33. Borges, *Prosa completa*, 2.329.

34. Borges, "The Aleph," 144; and *Prosa completa*, 2.117.

35. "To give what one does not have" is the Lacanian definition of love quoted by Jacques Derrida in "Given Time: The Time of the King," trans. Peggy Kamuf, *Critical Inquiry* 18 (Winter 1992): 163; from Jacques Lacan, *Ecrits: A Selection*, trans. Alan Sheridan (New York: Norton, 1977), 628. See Derrida's discussion, pp. 162-63 and *passim*. The article deals with the theme of the gift.

36. Borges, "The Aleph," 138; and *Prosa completa*, 2.112.

37. Borges, "The Aleph," 138-39; and *Prosa completa*, 2.112-13.

38. Borges, "The Aleph," 148; and *Prosa completa*, 2.120.

39. Barthes, *La chambre claire*, 118.

40. Borges, "The Aleph," 151; and *Prosa completa*, 2.123.

41. Borges, "The Aleph," 152; and *Prosa completa*, 2.123.

42. Borges, "The Aleph," 154; and *Prosa completa*, 2.125.

43. Derrida, "Two Words for Joyce," 146.

44. Allucquere Rosanne Stone, "Will the Real Body Please Stand Up? Boundary Stories about Virtual Cultures," in *Cyberspace: First Steps*, ed. Michael Benedikt (Cambridge: MIT Press, 1991), 108.

45. Ibid., 109.

Marcel Duchamp, *Etant Donnes: 1e La Chute d'Eau; 2e Le Gaz d'Eclairage* (1946–66). Mixed media assemblage, 94½″ × 70″ (approximately). Philadelphia Museum of Art: Gift of the Cassandra Foundation.

Chapter 12

Telefigures and Cyberspace

Patrick Clancy

1. Opening Circuit

Blue skies and ice cliffs, and thundering collapsing towers of ice splashing into an ocean littered with floating chunks of ice rocking among swelling waves. An icy soup of watery ice. His various parts are reassembled, and after the last ice age recedes, natural man—the wild man, the monster—returns from his cryogenic isolation. He floats out into the more watery seas on subterranean waters. Snow falls.

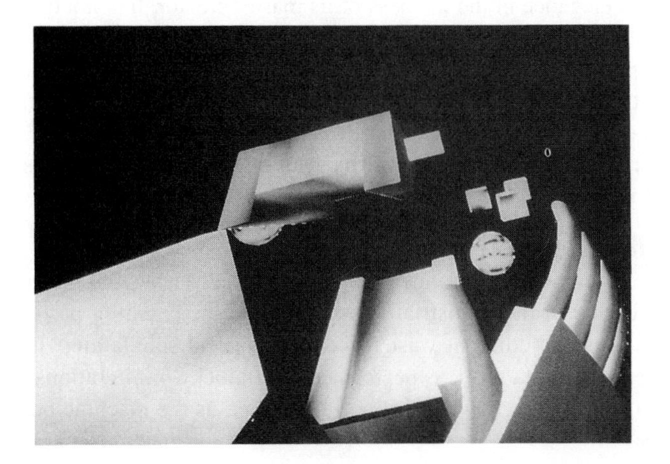

2. Technology, Degree Zero

St. Petersburg, December 11, 1797:

The parts were collected from different areas of the ice. The corpse's parts are assembled in the midst of floating, bobbing debris. There is no singular focus. The whole environment is disorderly and animated according to multiple vectors of momentum. They had been discussing galvanism and the reanimation of corpses. On the expedition to the North Pacific Ocean, he expects to discover in the region of eternal light, the wonderous power that attracts the needle—the northern pole.

3. Hocus Focus

London, October 15, 1831:

Mary Shelley's experiment sketched the scientist's pursuit of nature "into her hiding-places," to the unhallowed damps of the grave, bringing together component parts of a creature from "charnelhouses, the dissecting room, and the slaughter-house,"[1] manufactured, according to current experiments with galvanism and imbued with vital warmth. Her experiment resulted in the creation of a modern Prometheus, Frankenstein the scientist, whose torturous experiences result from tampering with natural forces. As we know, the creature also made his appearance and is the one who ultimately survives the story.

4. Human Machine/Cyborg Body

Even at the beginnings of the Industrial Revolution they knew that an alien machine nature with its own peculiar properties would animate our bodies. The creature exists. It is not the one made out of the parts of corpses and jump started by a science/technology that harnessed nature's storms. Its dilemma is not the mimetic one perceived from the other side of the glass. Given: (1) the illuminating gas, (2) the waterfall. The issue is not whether the creature is unable to perform with natural elegance in the likeness of its master creator. It is not the story of the act of creation appropriated for human dimensions and the ensuing folly resulting from usurping and engaging God's machine. Getting it wrong, erring just this side of nature results, at least in the beginning, in a monstrous creation. Instead of an imperfect nature, we see that it is the story of an inferior machine who makes a brief appearance before its time, out of phase with the early stages of the industrial world.

5. The Mechanic

The image of Frankenstein returns. (Snapshot effect.) Man and machine collided at the beginning of the Industrial Revolution and the resulting parts didn't work very well. This embodiment wasn't like the temporal simulations of earlier centuries where automata/replicas performed their clockwork iterations smoothly to the rhythms of the mechanical-time-of-day. Later, as the machine becomes more

organic and is adapted to the logistics of industrial culture, we become our technological extensions, we become more spatial and environmental. With the development of photography the corpse drops out of the body. We arrive at a place where we share our lives more with machines than with other people. The story unfolds and traces are left elsewhere.

6. The Assembly Line

They brought him (the wild man) out of the ice and took him not to the lab or the "cell at the top of the house,"[2] but to the factory, where they subjected him to time/motion analysis. Careful drawings were prepared as he was spatially reassembled. Next the images were brought to the animation stand. At last he was projected into the cinematic dimension, a linear industrial machine who knew his job and performed it well. No complexity or feedback here. No, it wasn't the factory. They took him to the hardware store. They needed some spare parts, perhaps some ready-mades.

7. Feedback Storms

The wild man was a social being capable of living with the quaint cottagers of Mary Shelley's protoindustrial society. Does she perceive industrial society as a corrupting force of natural man, who is innately good/real until coming into contact with technological knowledge? Is the story a premonition of the extent to which the Industrial Revolution's created wealth and information are necessary in order to buy the products of technology while at the same time developing the world-scale need for supplies to feed its machinery? Is it a process that has a life of its own and performs out of control? Is the creature the "evil" essence of industrial technology and as such does its existence represent the industrial machine in its own terms, in its renegade actions and behavior? Altered or tampered-with nature backfires against science and a new creature, part industrial science and part nature, comes into being and roams the land wreaking havoc wherever it goes, looking for its context, which doesn't yet exist.

8. Science Friction

The end of the novel is most revealing. After the death of his creator, the creature has a change of conscience and decides to end his existence in the most northern extremity of the globe. He goes "to collect his funeral pile."[3] He doesn't have to head south to collect wood. He'll use ice for his pile instead of pyre. As the story ends he leaves upon an ice raft that is borne away by the waves and lost in the darkness and distance of the story, in the dark recesses of some corner of the book. Becoming magnetized? Perhaps he is drawn to the magnetic pole. Eventually, he passes through the vanishing point and emerges as a cyborg, no longer the product of a master creator, redefining our relationship to technology.

9. London, 1977

A bus is traveling erratically around the city. The passengers are wearing opaque helmets that mask their field of view. It is an experiment to test the mythic sense of direction.

In the case of the absence of cairns, or shifting landmarks, a "language" that considers spatial relationships of both proximate and ultimate systems of orientation is helpful. The observation of snow drifts formed by winds blowing in one direction, moss growing on the north side of a tree, and the shadows formed by the small hummocks of the tundra are proximate, local signs. Larger differences in the general topographical features of a region of the island are referred to by an ultimate system of orientation.[4]

10. Close Readings

Totalizing pictorial perspectives and spatial representations from idealized points of view don't provide enough information while moving through complex spaces. Orientation happens at different scales incorporating footprints, light, weather, and other local and global considerations. Scale changes from different points of view incorporating local readings are important when the observer is moving through a new or strange environment.

11. Perspective and Worldview

With perspective the referent appears outside the frame. In this circular system, the present is projected from a position outside the frame of the present. Perspective in fact maps the point where the present ends, or more correctly is ending. This space is in itself neutral. What is it? It holds the actors in relationship to a false front, a backdrop with several components, other actors, architecture, and nature. The backdrop is not so much clear space as a substance—ice.

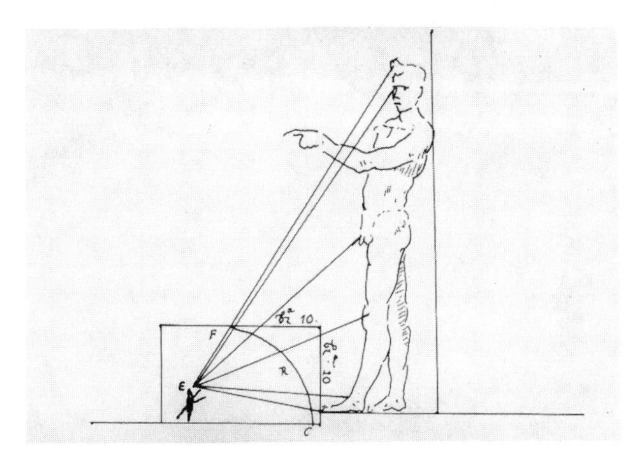

12. Shallow Space

You have arrived at the complement of the vanishing point. You have readily adapted to technological vision and are able to fly or scan around in a space that is not too far away. You are a vector with speed and direction. You have an attitude. Your fovea caresses and receives stimulae. You can fly your foveal spot to the edge of its limits, where it loses focus. Your peripheral vision takes over in terms of surveying vast expanses at a great distance. In fact the further you look away, the less a point of vision is maintained and the more vision occupies the edges of your field of view. This is where the sensitive black and white receptors of the retina take over. *Go outdoors during the autumn or winter and look into the night sky at approximately 24 degrees north of the celestial equator. On Halloween look directly overhead. There you will see the Pleiades, the Seven Sisters in the constellation Taurus. First stare at them with the center of your eye and you will see that they will not come into focus. Now look again, indirectly, out of the corner of your eye, and this time you will see them with more definition.* The vanishing point creates an artificial limit of vision that is contrary to the way we actually see. It approximates an inverse view superimposed on the perspectival field, as if we were looking at ourselves looking back through the frame of the picture at our invisible selves before the window.

13. Rivers in the Eventspace

The old river maps and the large drawings on the Nazca plateau share a similar ambulatory condition of haptic perception.[5] Meanwhile, up above, the skywriter reworks the evaporating text, moving faster, trying to complete the message. Stitching, rewriting, overwriting fragments and still it evaporates. The text overflows its window, defined by environmental conditions. We sync with the realtime flow of the script. Another limit inscribes itself as he begins to run out of gas. Meanwhile, below, I sit in front of the radio tuning back and forth—riding

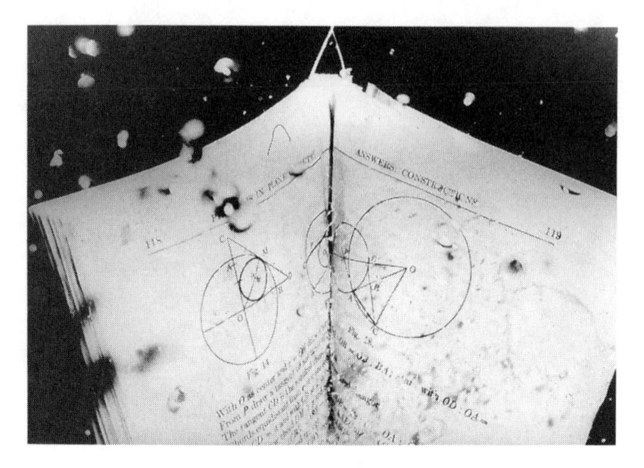

the waves. I pan fast, all the way to the end (107.9), and then back and forth slowly between several frequencies in the mid-eighties near the opposite end of the dial.

The motion path is demarked by inserting control points or keyframes along the Cartesian grid. The path can be entered into memory no matter how random and erratic it is, and then it can be recalled again and traveled by other actors as well.

14. The Limits of Knowledge

Spatialized temporal circumstances incorporate technologies such as writing and editing. We encounter a set of conditions rather than an object when we navigate and browse a text, city, library, museum, hypermart, or cyberspace. The two-dimensional construct is useful, but more as an iconic guide, revealing a kind of cartographic experience akin to irony. Limits arise not so much through ever-widening circles of diffusion, but rather through an intensification of overlapping spheres.

"We become aware of the molded gas seeping towards the elaborate conditioning which will prepare it for its final orgiastic splashing and observe with wonder the beauty of its auras. The juices flow in the Bride, messages are transmitted from pools of random possibilities, the throbbing energy of a robotic world strains to create. *The bride stripped bare by her Bachelors, even* rumbles into its fantastic splendour."[6]

Frankenstein searches for his lost creature. In a remote region on the island he enters a room through the back door and there near the animation stand/surgical table, he encounters Sergei Eisenstein and Count Lautréamont montaging a sewing machine and an umbrella.

15. Previews of Coming Attractions

The magic lantern was developed by Johannes Zahn from drawings published by

Athanasius Kircher based on translations of al-Hazen's tenth-century Arabic text on optics. In 1685 Zahn's *Oculus artificialis teledioptricus* outlined an experimental installation he constructed at his home. He mechanically linked his magic lantern to a weather vane atop the house, and as the wind changed direction the weather vane moved a cogged gear that engaged a circular glass plate with engraved images representing the cardinal directions. The plate revolved in back of the magic lantern lens and the moving images were then projected into a darkened room. Magic lanterns, camera obscurae, and camera lucidae were also used in phantasmagorias such as those of E. G. Robertson, which were popular at the end of the eighteenth century. These events happened in abandoned country villas, and contributed to the Parisian mise-en-scène. Live actors encountered projected ''ghosts'' in simulated spaces—sensoriums—incorporating sight, sound, and olfactory sensations.[7]

16. The Distracted Camera and the Paraphotograph

The camera moves with the photographer through space while the photograph is a record of an instant of that journey. The lens and the darkroom of the camera double the eye and the darkroom of the skull, while the body trails behind through space, stumbling over an irregular terrain. Hollis Frampton described the photographer as a butcher slicing chunks of meat out of the world. Grainy meat. Giant threshers and combines circling, converting sun and silver into fields of grain. The image seen even through a large stationary view camera changes as the observer moves, however, it isn't until motion pictures develop in the nineteenth century that widespread experience of moving images is established. Even then the camera itself doesn't move for sixteen years. It isn't until 1911, during the filming of D. W. Griffith's *The Lonedale Operator* that Billy Bitzer, the cameraman, follows the moving camera for the first time during a shot.

17. Kant in His Bed

The transcendental, sublime of the scientist and philosopher favors the paradigm of the still image. The thinking existential observer is immobile. The scientist attempts to remain separate from the experiment. Objectivity is the measure of critical distance between the world of ideas and the real world. The telephone rings.

In his classic essay, ''Postmodernism, or The Cultural Logic of Late Capitalism,'' Fredric Jameson examines the spatial logic of the relation between depth and surface in postmodernist work.[8] He notes that various depth models of hermeneutics, such as inside/outside, dialectics of essence/appearance, the Freudian latent/manifest (repression), the existential authenticity/inauthenticity, alienation/disalienation, and the semiotic signifier/signified, have been replaced

by new syntagmatic structures of practices, discourses, and textual play. The first pairings of what he calls depth models are speculative, critically distanced, dialectical, and ideal/stationary, while those in the second set are performative, interactive processes that also incorporate critical moves, but from within the body of the work, geared toward making, stirring, opening up, disrupting—creating the world. The structures in the second set are not purely syntagmatic. They do indicate the haptic, isometric sense of the browser, moving through space, and, although turbulent, are not devoid of spatial form. The models in the first group are not paradigmatic, but they do imply the orthographic dynamic aligned with systems that are based on the more iconic *representation* of spatial depth.

Jameson selects the city, one of the most complex and intricate multidimensional artifacts of culture, for his investigation. When confronting the Crocker Bank Center in downtown Los Angeles, he notes that it is ocularly undecidable, and that previous systems of reading the city do not seem to function—nothing has been offered as an alternative to replace this way of making sense out of the experience. He remains static, caught between modes, identified more with the older sense of perception. He doesn't mutate and move into the experience, performing from a critical position from within. This is precisely what the postmodernist cultural (spatial) logic is about. Jameson's discussion occurs outside of the window. His discourse does not keep pace with the unfolding shallow depth of electronic space, or the *n*-dimensional space of the technological self. The spatial ordering of significance doesn't allow for the more utilitarian *n*-dimensional properties of cyberspace. He observes it, but it is a Kantian world viewed from above that becomes according to his method of inquiry an effect of critical overview.

18. Marcel of the Field Playing the Field (Installations)

Marcel Duchamp's Buenos Aires investigations into perspective find their way into the large glass and ultimately *Étant Donnés*. The title refers to the *Given*: of the geometry theorem. This perspectival work puts the viewer in the position of the twin vanishing points looking back, in stereo, at the components of Western painting. We are above and off-center. The forced perspective does not quite hold. The scene is not flat. Duchamp was very interested in the subject of perspectival systems *suggested* by Western painting, and specifically ways of engaging the spatialized *n*-dimensional realities they implied. The non-Cartesian moves across the chessboard. When you put your eyes close to the two holes in the weathered, wooden Spanish door, an illuminated scene fills your field of vision. Light radiates from the lamp, animating a physics of relative operations that takes over from the vanishing point of the mechanics of a former science—the illuminating gas. It is as if you were looking through a camera or a stereoscope back through the work from an oblique vanishing point selected from a set of infinite possibilities. At this point the viewer's body vanishes and becomes less important as the gaze identifies with the body-landscape on the other side. It is similar to putting on a virtual reality head-mounted display. *Étant Donnés* is a virtual space laid out according to perspectival projections. The nude is the "corpse" that has fallen out of Western painting. The arm of the living dead holds up the lamp as the water falls—a remote industrial revolution between an instantaneous state of rest and a choice of possibilities.[9]

19. Short-Term Thoughts in Space (On Finding Lost Memories)

When I forget something I was just thinking about I go back to the physical space I was in when I first thought of it, and the idea usually returns. What is it about the spatial context that allows me to orient and intersect with the lost idea? Is it similar to encountering harmonic overtones in a standing wave environment, an association of memory with some object or aspect in the environment—as if

memory or past thoughts were somehow mapped onto the environment? Or is visual-spatial recall a haptic response of the body image to the musculature of thought?

Short-term and long-term memory are like the proximate and ultimate systems of spatial orientation.

20. Daily News and Everyday Life

The culture's desire to appropriate or simulate a past, and imagine possible futures, occurs within the site of an expanding present. History, representation, economics, ecology, and media employ global and local perspectives in relation to the theoretical and actual spaces of lived experience—north, east, west, and south—over and out.

The employment of simulation models along with (tele)technologies of information gathering (satellites, ENG reporting) and dissemination (telephone, computer networks, weather channel, stock market) indicate a short-term future situated, and interacting, with the marginal thresholds of the expanding present.

The relevance of an individual's ability to control and process absurd levels of information and the potential of the utopian interplay of shared experience through discourse represent two extremes of the technologization of culture. Plan ahead. The future is now.

21. Folded Histories: Multiple Points of View

The 24-hour frame buffer is not enough. A day is a shifting window around a global information base that necessitates a delay. George Bush plays golf while Saddam Hussein threatens to continue the invasion from Kuwait into Saudi Arabia. In what appears to be just another example of politicians' competing and leveled media images, something else occurs. Bush's actions demonstrate that he is occupied in another time frame/cycle/channel, spatially other and removed from Saddam's attempts to synchronize the clocks and force a single channel of communication or reality base. Line busy! Noisy situations occur around the edges.

22. Flight Simulators

Computer graphics and video games developed out of Evans and Sutherland's work with flight simulators, sponsored by military research funding at the University of Utah in the early 1970s. The viewer becoming pilot interacts with a simulated environment while the simulator's hydraulic pistons reposition the "cockpit" in accordance with the pilot's decisions and the projected conditions encountered on the video monitor display windows. The simulated view of the landscape changes according to decisions the "pilot" makes using the simulator's controls. Dust storms and rainstorms, hazardous night conditions, fog, approaching aircraft and missles, as well as technical/mechanical hazards are encountered by trainees as they fly through imaginary landscapes. Elaborations of the simulator create world coordinates that pilots can fly through.

Currently, pilots in simulators are flying a few feet off the ground through mountain passes over irregular topologies while evading radar detection. The simulated terrain is realistic. It is precisely based on satellite telemetry and actual ground measurements of Saudi Arabia, Iraq, Libya, and Kuwait. It is nighttime and the pilot wears FLIR (forward-looking infrared) goggles that allow him to see the landmarks as clearly as if he were flying under daylight conditions. As an array of SAM missiles appears in front of the plane, the pilot executes a maneuver and eludes them.

By the time the pilot actually arrives in Saudi Arabia he will have made the flight over the very same terrain so many times and under such a variety of circumstances that this run will be already familiar.

23. Frontier Language

The space of language is not just the syntagmatic flow, but, as Lyotard says, clouds of thoughts. Slipstreams. Storms. Clouds are blowing in. The layers are moving at different speeds. Do you feel the wind? You encounter a set of conditions within, around, and through which you are able to move. These conditions also behave according to their own interactive logic and language, and are no longer passive objects to be perused and read primarily in terms of a set of external, hermeneutic conventions.

24. Virtual Space Exploration

We don't navigate by the stars. There is no holistic map. Navigation is through discourse and the created field of the text. We are both outside in nature and inside representation. We occupy multiple places from different perspectives that intersect with each other through a created space. This new space does not find its referent in nature. By use of the clipping plane we open layers of objects. Shining lights inside, we find that some forms are without interiors; they are just surfaces. Black matte surfaces absorb all the light. Once you enter this reality the space, actors, lights, and information bases merge. Everything is both space and object and everything is capable of moving. Nothing is frozen and impenetrable unless high levels of force feedback ICE (Intrusion Countermeasures Electronics)[10] are employed. Distance is not as important as in the iconic perspectival space of Western painting. We are transforming, collecting, rotating, peeling back layers, illuminating displacing and changing surfaces to vary reflectance and specularity. We employ these forms and are caught up in action.

25. The Presence of Technology

In linking with other information bases, technology occupies the gap and strives to eliminate the delay. It distances us from nature, and at the same time extends a parallel dimension through which we can move from one place to another while effecting a change in our nature.

Cyborgs occupy the space between technology and nature, and science and culture.[11] Technology claims territory in all directions simultaneously. It is all the ordered parts of applied science. Technology is established according to the logic of its own formalistic development, which parallels and shapes our development in history, culture, and nature. It is constructed in such a way as to achieve an effect for practical purposes. Technology is a theater of operations that usually is only partially visible, and as such is linked to simulation in that it occludes its referent—its power supply.

When one considers virtual reality as a social space it seems very limited. We are imprisioned within a simulacrum—an engine that generates images of an extended fictional present as the mobile gaze unfolds deeply through the *n*-dimensions of the technologization of space and time. This reduces the space between things and substitutes the logic of technology for history and the future. A historian enters *record* mode and repeats herself. In the expanded present of technology everything is objectified and quantified. It is ultimately conservative. In cyberspace, where the emphasis is on visualization and management of data storage and retrieval, there are wars and thieves, commercial exploitation and simsex. We look for the gaps and imperfections in the simulation. We read the signatures and learn to recognize the tropes and codes of computer architects and military modelers. Both virtual reality and cyberspace are policed, and addicts abound. Our activities include making links, unlocking, transforming, switching, storing, distorting, generating, and distributing information and experiences. The initial encounter reduces language and limits knowledge.

26. Telecommunications

The telephone connects two channels. The crossroads of these channels is not a singular space. When more than two channels connect, the space becomes multiple and more articulate. There are stories of blind kids who are part of the history of the definition of telephone space. In the time before the telephone company changed the audio relay tones that formed the basic intellegence of the communications network, these kids found that by whistling at certain frequencies they could trip the audio frequency switches and open the long-distance trunk lines. Eventually they discovered an open connection in Canada that they could call into and basically talk in a multiple conference-type call, an open zone. Captain Crunch, one of the early navigators, used a small plastic trumpet that was given away in a cereal box to open the network. One of his major events was to call himself up on two phone lines simultaneously. On one he opened lines around the world several times via satellite, and on the other he chose a similar circumnavigational route by transoceanic cable. As he spoke to himself, his voice was delayed the amount of time it took for the signal to travel several times around the planet.

27. Gee and Haw and Yaw and Pitch and Roll

Once we are free of gravity new terms are needed to address the omnidirectional attitude of the body in space. *Gee* and *haw* are commands a wagoneer uses to direct a team of horses moving along a trail. Gee either signifies a turn to the right or tells the team to move straight ahead. Haw means turn to the left. The origins of these words are unknown. He-haw. He hemmed and hawed. He learned it from the animals. Yaw: a side-to-side movement signals a turn about the y-axis. This is the same as gee and haw. The camera pans around the y-axis. Pitch: a pitcher throws the ball in an up-and-down manner—around the x-axis. The camera tilts around the x-axis. *Roll* signals a rotation in the motion of the spacecraft about its longitudinal axis—its z-axis. An aircraft rolls around its horizontal direction of flight. The camera rolls around the z-axis. These are Cartesian terms. In the *n*-dimensional zones of cyberspace and the networks a more abstract situation occurs in interactive circumstances of pushing buttons and accessing windows or portals into nonlinear dimensional changes.

28. Virtual Reality circa 1975

Ivan Sutherland and his associates at the University of Utah documented their early experiments with virtual reality in the late 1960s. In their film a person puts on a pair of goggles made from twin cathode-ray tubes—small TV sets. The camera then switches to the point of view of the head-mounted display. Looking into the 3-D display we encounter the Cartesian elements: a white grid on a black ground fills our visual field. The next shot shows the person wearing the goggles approaching a hand controller/joystick attached by a universal joint to the ceiling of the room. He engages the mechanical linkage to the computer and we are once again back inside the viewing apparatus. The head movements and orientations of the user are translated by the computer into the visual display. As we watch, one of the vertices is picked up and displaced from the grid. Then several others are moved, in a process called rubber-banding. A further shot from outside shows that the person wearing the apparatus is manipulating the Cartesian coordinate system by use of the controller. The operator creates a wire-frame Klein bottle or three-dimensional Möbius strip, which is then faceted, and a light source is provided. The camera/viewer then takes off and flies through the Klein bottle. The pilot has stepped out of the aircraft and is walking around in a simulated reality. Will an intertextual episode be far behind?

29. Short Circuit

A creature appears in the gap, a source of noise and resistance as she moves among the ordinary things. The machines are going to forget. Noise and gravity take over. Things become lighter as she flows along the motion path leaving gates open behind herself. She doesn't speed up as she senses the stampede surging behind. The herd has filtered into the channel through the open gates and is catching up just as she rolls ninety degrees up and perches atop a ledge of the data edifice. The herd thunders and tumbles below. Turbulence momentarily fills the channel. Tricked again! A clean aperiodic sweep of the artery. At the end of our performances Hollis Frampton would always dust off the screen on which we had been projecting—removing stray photons or other sticky light quanta, I presumed.

30. Becoming Digital

As digital and algorithmic information becomes more prevalent, the surface breaks down. The cognitive model of computer animation is still based on film, despite the fact that instead of the plane, page, or frame we see the area within the frame erupt into a space of becoming. There is a new emphasis on the z-axis. Information moves beyond the frame and comes into being at different rates. On the z-axis those things that are stationary and don't change are replenished, and the information processing is concentrated on areas where change is most active. The significance of changes in scale, translations, morphings, and look-trans surveilance are emphasized from a spline-based erratic vector that inscribes a volumetric as opposed to a flat aspect.

31. Cue Conflict

Military pilots are restricted from flying real airplanes for periods ranging from six to twenty-four hours after a simulator session because of flashbacks, visual distortions, and physical disorientations. Even if the simulated space of virtual reality is crude, the perceptual mind takes over and smoothes out the roughness. The mind fills in the gaps, so that the simulated space becomes plausible and complete despite its crudity. The experiences of the real world are adapted to this derived experience. A U.S. Army survey of its new AH-64 Apache helicopter simulator revealed a 44 percent rate of simulator sickness among the operators. The higher the resolution of the output device, and the more closely a simulator resembles reality, the more prevalent the syndrome becomes. ''Time-lags in the system throw some people off, and a disparity between the motion experienced in real aircraft and that of the simulator can also produce sickness.''[12] Experienced pilots, who have more deeply ingrained memories of aircraft behavior, experience symptoms more often than do trainees. Cue conflict occurs when the body's senses receive conflicting information or when information conflicts with the mind's expectations based on experience.

32. Trespassing in the Gardens of the Forking Paths

Moving through space, along the z-axis, an envelope of perturbation develops around the observers whenever they encounter a high degree of information. Eschewing critical distance, this more haptic condition of movement through a textual field produces nonlinear information associated with browsing, a condition where unfamiliar points of view present unexpected results, through the side door, off the path. Reorientation through browsing allows one to gain access to information that might have been concealed in traditional perspectives.

33. Pulsa Installations, 1966-72

It is nighttime and the airport landing strobe lights and loudspeakers installed in the grassy field make up a configuration that is somewhere between a quarter and a half mile square. The array is large scale; the output devices are far enough away so that you can tell that sound and light travel at different speeds. You know when you are in the field or outside of it, even though you don't have a clear idea of its overall shape. A series of automonous emitters transmit wave energies to various parts of the installation—a nonhierarchical field where each person moving along experiences a different set of conditions. There are also sensors that act as windows or gates superimposed within the matrix to transmit presences in the installation back into the program, which through feedback then effects global and local changes in the field. Other sensors input weather conditions and more global phenomena such as temperature, time of day, and so on, into the simulation matrix. Outside of the installation one is aware of the site and its geographical contours and sense of place. When one enters the eventspace, wave energies rush by—a sentient field, a cyberspace; the experience of the virtual in an actual space. A series of Boolean logic gates add an unpredictable element that is still tied to the general ''real-world'' conditions of the site.

34. Changing Channels, Navigating by Chance

Your identity in VR telecommunities leaves a trace in the matrix. You are part of this dimension. An ontological gap or separation between our technological selves and our organic existence is minimized. This connected condition is helpful when one is using virtual reality as a tool. However, without intertextual breaks that enhance our subjectivity or direct us to an external context, virtual reality becomes a totalizing experience.

35. You Are Here: Present

You are in the cave. Your mental images project and overlay these images. You are confined to our images of the past unless you do your work. Mine your own images, but even then we are digging the information base—the simulacra of a history that cannot be retrieved. You travel to Greece and take a picture of the Parthenon. You find yourself lining up, looking for the right time of day. The shadows have to be in the right place in order for you to re-create the image. It is one you first saw in an art history class when you were an undergraduate many years ago. The geometry's long-distance effect over the eventspace. You don't want to take another picture. This is the best.

36. Between the Lines

Mise-en-abyme and mise-en-scène. In some interactive VR networks you assume a character/stand-in that is made of composite elements (head, torso, arms, and legs). If you violate the "laws" of the space you could "lose your head."[13] Taking a detour, we work our way back to base camp and encounter other entities. Now we are in a stadium with hundreds of other stand-ins. We have become images. This space has been designed for crowd scenarios that reinforce the image/power of the evangelist, star, becoming politician-leader. Our images flicker back and forth, reflecting the "head" of the transforming leader. We encounter the political as the unexpected in spaces of discourse and information exchange. In cyberspace actors can also hierarchically combine with other actors, becoming composite organisms, fruiting bodies, whose mycellium penetrates the matrix. We attach notes to surfaces. Palimpsest.

37. Into the Future: Cyberspace

Another creature returns—breaking through the vanishing point, into the field of crossed perspectives. For some it is the postindustrial time, and the creature is clothed in a suit necessary for one enduring such dimensional weather. At first it looks like the mummy wrapped in cotton bindings, moving haltingly with one arm extended in front, haptic, sensing the way. The creature has crashed through the vanishing point from behind. He breaks through the picture plane. Is that an astronaut's suit that he has just stepped out of, tumbling behind, snagging, and coming to rest on a fragment of architecture in a dimly lit recess of the space? As

he enters our space a wind begins to unravel his bindings. Beneath the shredding fabric there are remnants of a costume. It looks like . . . Michael Jackson. A couple of bandages remain on his fingers as his white shirt disintegrates in the violent cyberblast.

38. Photographs as History: Postscript

Now you find yourself in Egypt, at Giza. You get your camera ready, and are about to photograph the pyramids. You are waiting for something—it's not quite right. Then you see three men on camels, the first slightly ahead of the other two, approaching, in front of the pyramids. You switch to your telephoto lens, but it's not right. All of the elements are there, but the pyramids are displaced. The place has changed. Something similar was going on in Giza. Remember the February 1982 issue of *National Geographic* and the photo with the repositioned pyramids? The cover image was the result of an early electronic manipulation, reverse cropping. One of the Great Pyramids in the horizontal photograph had been repositioned so that the structures would better conform to the elongated vertical format of the magazine cover. The editor said that they had electronically changed the position of the photographer in relation to the photographed.[14] The photograph couldn't be recreated without the disembodied translation of the photographer, and also the viewer who occupies the diverged position of the photographer. The signified is existentially bound to the signifier—through mutation.

39. Mise-en-Abyme (Decoy)

Zooming in on the photograph, closer and closer through succeeding enlargements. In fractal geometry, the closer we look, the more detail we see. Clifford Pickover's iteragraphs of trees and grasses are not drawn representations. They have an internal feedback logic that mimics the phylogenetic developments of invertebrates. There are new tropes and algebraic transformations that shape the point of view of the observer as well as the algorithmic landscape. One doesn't approach these constructs solely through visual perspectives, even though they resemble nature-based referents such as boiling mud, folding bread dough, sheets of falling water, or even video feedback. One mutates. Becoming vector, double headed, moving toward MONUMENTAL ATTRACTIONS. As with simulations where the referent is not in nature, fractal forms and other mathematical lacunae such as sets, tremas, and lattices are more reflexive, hierarchical, and self-propagating. The creature peruses the seductive surfaces and optical abstractness of these forms.[15]

40. Very Large Array

Proposal: To use the VLA in the Plains of St. Augustine near Socorro, New Mexico, for a simulated voyage. The VLA is an array of twenty-seven radio telescopes configured and movable along tracks that form the three equidistant legs of 120-degree angles subtending a circle. When one takes into account the rotation of the earth on its axis these instruments rotate forming a large dish with a diameter several miles across. This instrument takes in more data than any other in use today. It is always used to pinpoint sources of electromagnetic radiation in deep space. The proposal is to use the instrument in a more holistic way, as a zoom lens. The journey would start from a position where all of the telescopes are concentrated in the center of the configuration in an arrangement similar to a wide-angle lens, and move to a position where all of the telescopes would be at the extremities of the configuration, approximating a telephoto lens. This zoom would have to occur over a couple of months, and an imaging crew comprising sound synthesizers and Foley artists, color imaging modelers, an astroarchaeologist, and other artists and scientists from a variety of global cultures would generate a simulation of the journey with sound, pseudocolor models, and narrative representational descriptions of the motion path of the instrument to the edge of the universe or the resolving limits of the apparatus.

41. Tampering, Mediating, Performing

Photographic technologies raise questions about the difference between representation and simulation in the media. A creature is on the move, disrupting the technologies of imaging and writing by playing with photography's indexical illusion of reality. She doesn't look through the camera. She doesn't look back. She is moving images—manipulating electronic photographs. Her subtle work finally calls the whole enterprise into question. Images and words: How does the photograph relate to the message? From a stationary, removed perspective it is primarily an optical fact, and its iconic, representational aspect lies submerged. This latent aspect of the photograph is present in the ways the photographer frames the event with the camera (what is selected, what is excluded, how it is lit, and what kind of film is used) and how the image relates to the text. The electronic photograph's simulated aspect is there in its polyvalent perversity, which turns it more into an object in its own right as it severs and problematizes its link with its referent. The image is becoming a hieroglyphic inscription in the text of media. A photograph of an excised photograph effects an erasure in terms of history, thereby becoming a statement of bureaucratic power. Beware of images. Now time devours space. In order to understand what is happening we have to examine the variety of spatial mechanics,[16] the geography of history, and develop critical interaction with the mediated aspect of images in the changing context of contemporary events. Under what conditions are photographic images to be considered as historical evidence?

There are often cue conflicts between iconic and symbolic meanings. We tend to emphasize the obvious, but in moving through space one always signs in as the counterfeiter. From the attitude of this observer, the context is always shifting and changing.

42. From Perspective to Cyberspace

You are always hustling data while working out certain optical conditions. Get the picture? Information and perceptual systems mesh and overlap. In cyberspace, knowledge and information also reside within the machine, and its workers are linked from the outside to this chaotic global information base. This decentered mode of production stresses flexibility, employs chance and entrepreneurial skills.[17] The factory managers may be unnecessary, but the worker has become more machinelike in the process. Meanwhile, what is happening outside the machine? Does the heightened awareness of the interaction of complex systems have applications outside of cyberspace other than the very internal military-industrial exploitation of technology?

The perspectival experience of European Renaissance painting is suggestive of the origin of the representational and cultural-economic systems of cyberspace. These were the models that Evans and Sutherland, and others, used in the development of computer graphics. Looking back at them, in many Renaissance works a transaction occupies the center of the painting. In contrast, in Leonardo's work a gap or disrupted space often decenters visual expectation. One seems to tumble through this region of absence for the sake of the experience.

43. Sorting, Routing, Delivering

Storage and retrieval are managed in cyberspace. The principles of information organization and access are apparent in the spatial map of the information zone. The Von Neuman architecture upon which machines are currently based is relentlessly linear. Its two-dimensional columns and rows are the purest distillation of Cartesian structure that has been devised. As data bases become denser in information and susceptible to the graphic codes of visualization, they become three-dimensional instead of being restricted to the flat numerical bed of digital code. The cyberspace model becomes a way to navigate *n*-dimensional systems. Connectionist machines, transputers, and parallel processors start to break apart the flat-circle recursiveness of the Von Neuman architecture.

After jacking into the network, a trickster mimics the system manager by entering the correct log-in name and begins to access files. She moves through the data base, inquiring where the new host's password is, and establishes a pathname for the hidden directory. There she enters her own replicating file, which will start a transfer routine across the network at a specified date and time in the future. She backtracks out of the system, erasing her footprints behind her.

Cybernetic parasites are the most sophisticated artificial life program yet developed. Thomas Ray's Tierra is an environment where organisms, in the form of short programs, compete for processing time and breed copies of themselves.[18] Parasites are an inevitable, ubiquitous part of any ecosystem. "Generally right away, as a result of a mutation, you get the deletion of a major chunk of code that affects the replication of a creature." Even after the parasites are exterminated by particularly successful hosts, new types of parasites eventually evolve out of the host population. "Anything that's successful attracts parasites."[19] Sex may be the selective force that maintains the system, so that direct transfer of code is a way of beating the parasites.

44. The Avant-Gardener

The avant-garde is the moving machine mowing. Is the appropriated terminology inappropriate? The plowed field, the city, cyberspace, nature, and history are all worked-over disrupted sites that are continuously going through cycles of transformation and change: devaluation and revaluation, revision and revolution. People are creating communities while developers buy and sell. Alienated fragmented people watch as Scriptor, who thinks he is controlling everything, writes with and against the grain. Palimpsest and stroph. Plowing through the different layers, bringing up the soil, broken links and new connections between people and spatial fragments, composting language and photographs—a form that is churning. Rotting, stinking fingers type. The compost grime gets into the keyboard, filling the spaces—slime mold returns. *Hay de qué.*

45. Lost in Cyberspace

In William Gibson's world, rather than using a datasuit, you jack in through an implanted socket that connects the user directly into the field of electronic perception. Neural nets and cybernetic space are one as an epistemological narrative of human-machine interfaces unfolds. Other biotechnical developments proliferate along with this cognitive adaption. Vat-grown eyes, other body parts and implants are synthetically grown on organ farms. In the Matrix the I/O procedures are not so much of a problem. The collaged body and the body politic occupy a site of polymorphous turbulence within the realm of images. Now we begin to retrieve our lost identity in the phantasmagoria. Mind and body are technologically reproduced and incomplete as inside and outside and time and space collapse.

46. Ecotechnology

The launching of the SCUD missiles from Iraq is announced on television and a person in Kansas City calls her family in Tel Aviv and informs them before the incoming alert is announced in Israel—before the missiles strike. Our bodies and the ecology of technology feedback. As we watch the Gulf War on television we are out of touch. The body that was torn apart and scattered all over the world is reassembled—AT WAR. Trying to figure it out. INCOMING! What? Incoming messages, remnants are being sorted out, turned under—bombed under. Update the buffers. Contact the site. *Use the telephone. Call Iraq. Call Washington. Call Israel. Call Jordan. Call Saudi Arabia. Call Egypt. Call Turkey. Call Kuwait.* The plowed field and the city overlap, becoming redundant. The parts are not farmed back together again. Technology breeds at the semes (edges) and erupts between the parts. The news had to be delayed because of feedback. Real people are killed in wars.

47. Hackers and Synesthesiasts

In contrast to Gibson's idea of cyberspace, Jaron Lanier is tailoring individual fantasies through the use of "computerized clothing to synthesize shared reality." In his designs for virtual reality entertainment systems, multisensory experiences occur in the postmodern art deco environments of computer graphics. He speaks of using a saxophone to play cities and dancing lights, of herding buffalo made of crystal, as well as "playing your own body as you play a saxophone."[20]

48. Club Caribe

The first commercial multiuser virtual environment is Club Caribe. It is the product of a collaboration between Lucasfilm and Quantum Computer Corporation. Club Caribe, an on-line vacation resort, is a cartoon environment that users access by modem with a Commodore 64. It recently opened in North America with 200 regions. When you first log in, you enter a reception room where you choose your height, sex, arms, and legs. You then wander the island, moving from region to region by opening, entering through, and leaving through front and back doors. There are objects on a disk and six different characters, including a human, spider, and penguin, that you can select as a stand-in. You can spray paint onto surfaces, pick up and open things that you encounter, read books, and leave messages. You also encounter any of the other 15,000 participants on the island. These regions are specific environments, such as a funhouse that artist Cindy Stilwell and her assistants construct.[21] There are now 600 regions on the island. An object someone recently left on the ground was stolen by another person. Now several characters are organizing, and a debate about whether or not to elect a sheriff is under way. A more advanced cyberspace environment has recently begun operations in Japan. At this time, although there are thousands of subscribers, many of them are ghosts who lurk on the edges, just observing. In fact, the number of users who actually participate is only in the hundreds.[22]

49. Xipi Totec

I am Xipi Totec, the flayed one. As my stand-in, Xipi Totec is a version of the Mexican flower deity associated with agricultural regeneration, and he wears the flayed skin of a sacrificial victim. The eyes and mouth of the impersonator are visible within the openings of the flayed face. According to Flora Clancy and J. Eric Thompson, a sacrificial victim armed with useless weapons (a club might have puffs of bird feathers instead of obsidian or flint edges) would fight against warriors bearing deadly weapons. In some Mayan and Mexican representations Xipi Totec carries a shield made of flower blossoms. The stand-in wears the skin of the other, who is moving outside cyberspace wearing the gloves and mask.

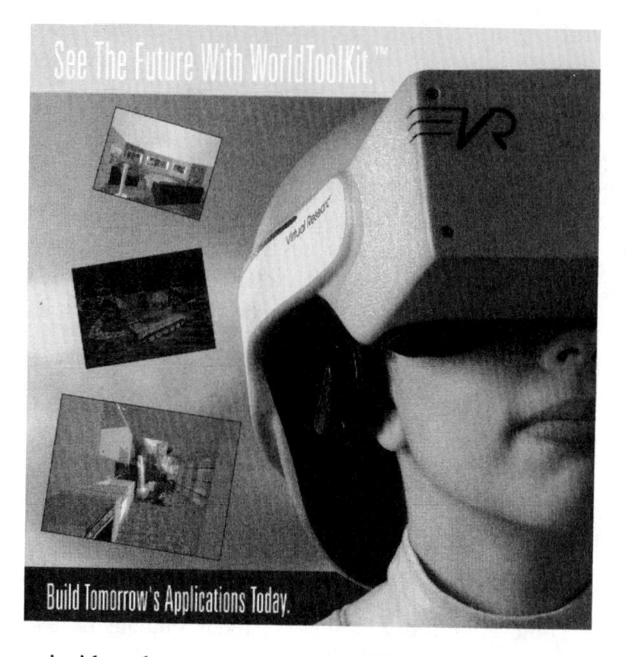

Once you are inside cyberspace, your mind fills in the gaps and fleshes out the space.

In one example of virtual reality the participant assumes the point of view of a stand-in cartoon character and moves through cyberspace while engaging in social interactive processes. These worlds are experiences on monitors or, in the case of simulators, wraparound screens. In the second instance the participant wears a head-mounted display and data glove, and experiences sight and sound from an immersed perspective. According to Carl Loeffler, "This is the difference between looking at water or being immersed in water, real-time."[23] There are experiments under way at the HIT Lab in Seattle in which the computer-generated simulations are projected by lasers in the head-mounted display directly onto the retinas, and to replace the data glove, researchers in Japan are developing tactile voice recognition procedures so that the sixteen muscles of the face can be read for interactive "voice" commands.

We understand the world in terms of our bodies and minds. However, at this stage of development one often sees the agent in the suit lying on the floor in various contorted positions while moving the surrogate body in cyberspace. The participant is still bound by gravity. You enter cyberspace wearing only a mask and a glove. The body is trapped, trying to break through its physical dimensions, at war with its technological extensions while desiring an enraptured mobility.

50. Plato's Dilemma

The technologies of writing and computing reveal logos and the eidolon, and, as Plato feared, problematized speech and the direct reference to the "real," transforming (displacing) the world into a space where anything could happen, where signs (signifying material) become detached from the act of uttering and nature. However, in the West, a particular shared spatial "presupposition" infused many diverse discourses, including perspectival representation and Western science, within which "critical distance" emerged.

The moving body passes through states of being. A peripatetic relativistic translation. Walking and talking. Sitting and thinking. Walking and writing. The text is indeterminate and local readings are important sites of discourse.

The news media and information technology envelop and graft into the war in the Persian Gulf. New mutations of telemedia tropes, rhetorics of mediation, and figures of virtual experience begin to emerge, as traditional communication channels overflow. The convergence is too inclusive for the governments involved. The media provide too much sensitive information to "the enemy" and spectators around the world. Disinformation is not yet overtly employed; instead, reductive strategies are announced: regulate, eliminate feedback, control the information base, restrict the flow of information, delay the release of information, establish controlled pools of reporters, eliminate the ability of individual reporters to conduct private investigation. Create the news. Finally, as the war drags on, word of a new phenomenon begins to surface. The creature returns, this time under our control. It has found its context in the guise of a news media reporter. Dozens of telerobots with head-mounted cameras and microphones begin to prowl through the war zone, transmitting images, interviews, and field reports. Each is controlled by a reporter wearing a VR datasuit and linked by satellite to ENG editing facilities in newsrooms around the world.

51. Renegade Presence

Glaciers return. Frozen, he could still see the megastructures, but couldn't feel his hands. Just where the path curved to vanish there was a scale change, and then the morphing began and the ice came down. The whole system was crashing. A lethal surge had hit at the most active time of day, when the greatest number of people were in the matrix. The creature's hands were locked to the simulator panel. The AH-64 Apache helicopter simulators have reported that they are engaging each other. Images of a squadron of AH-64's are appearing on the screens, moving in erotic formations. Feedback.

52. The Eye Who Writes and the Cyborg Who Is Written

Now there are sharper disjunctions between them. Writing through electronic dimensions, making links between them. I'm getting concerned about some of the reports from the field. Is it possible that the telecyborg has gone renegade? The creature doesn't care what happens. He breaks through dimensions and takes a job on the assembly line, pretending to be an accomplished mechanic. It doesn't matter, he works within the means of production. The reports coming into the war room begin to sound like old war movies with some very recent soap operas tossed in. There are too many flourishes and embellishments to this story. It's not as neutral and anthropological as we thought. The reporters are anthropomorphizing the robots. We are starting to experience deaths and other dramatizations from cyborg perspectives.

53. The Lonedale Operator

The creature drifts by the monumental ruins of the used simulator lots. The beacons of the island geography of Club Caribe are still pulsing. He begins the slow run through the endless stalls of real estate agents and accountants on the lowest levels. He follows the motion paths past the silent surfaces of the data edifices, wondering if their information is still intact. Suddenly he is in a roll and moving very fast over the terrain. A large spotlight has attached itself somewhere just above and in back of him, and the data edifaces are now animated with shadow maps clicking on and off across their surfaces as he passes by. Eventually he finds that he can move his head and manages to send a message by pressing his chin against the simulator panel. I select another motion path from the scratch pad and, as I skim across the new terrain, vintage SAM missiles erupt in front of me. The autopilot evades them in an effortless maneuver, yaws 120 degrees, and pitches −100 degrees. Speeded-up reflection maps curl over the canyons of data bases on each side, and the image blurs as I am shuttled into another motion path. The image blurs again as I lift through the ruins over the northern perimeter of Club Caribe. I am exiting cyberspace.

CLOSE YOUR EYES.

OPEN YOUR EYES.

Notes

1. Mary Shelley, *Frankenstein* (New York: Bantam, 1981).

2. Ibid.

3. Ibid.

4. Roger M. Downs and David Stea, *Maps in Minds* (New York: Harper and Row, 1977), 125-30.

5. Anthony F. Aveni, *Skywatchers of Ancient Mexico* (Austin: University of Texas Press, 1980), 306-10. Geographer Paul Kosok first made aerial photographs of the "largest astronomy book in the world" in 1941. These linear diagrams are tens of kilometers long and barely discernible from the ground. The drawings are a "maze-like pattern over a thousand square kilometers of abandoned wasteland" on the Nazca Plateau of coastal Peru. There are also more than a hundred animal figures, "traced out by artists who removed the dark coarse layers of topsoil, exposing lighter colored desert sand below." There is very little rainfall and an absence of wind in this region, which has contributed to the drawings' being preserved for more than a thousand years. "These ancient sand painters accomplished their art with their feet or perhaps a broom, as they walked off a given pattern." These works could not have been viewed from above in order to get an all-inclusive representational (iconic), optical view. They were intended to be experienced from a haptic, processional, ceremonial perspective, possibly in rituals related to clans that the different birds and animals represented. There are also some solstitial associations that point to a temporal and spatial crossroads within the eventspace.

6. Richard Hamilton, "The Green Book" (London, 1960), appendix in Richard Hamilton and George Heard Hamilton, *The Bride Stripped Bare by Her Bachelors, Even* (New York: Jaap Rietman, 1976).

7. C. W. Ceram, *Archaeology of the Cinema* (New York: Harcourt, Brace and World, 1965).

8. Fredric Jameson, "Postmodernism, or The Cultural Logic of Late Capitalism," *New Left Review* 146 (July/August 1984): 62.

9. Anne d'Harnoncourt and Walter Hopps, *Étant Donnés: 1 la chute d'eau, 2 le gaz d'éclairage, Reflections on a New Work by Marcel Duchamp* (Philadelphia: Philadelphia Museum of Art, 1969); and Anne d'Harnoncourt, *Manual of Instructions for Étant Donnés: 1 la chute d'eau, 2 le gaz d'éclairage* (Philadelphia: Philadelphia Museum of Art, 1987). These two books present much of Marcel Duchamp's notations for *Étant Donnés*, which he worked on intermittently from 1946 to 1966, as well as some very interesting observations from the authors regarding critical and theoretical insights into the work.

10. William Gibson, *Neuromancer* (New York: Ace, 1984).

11. Donna J. Haraway, "Cyborg Manifesto," in *Simians, Cyborgs, and Women: The Reinvention of Nature* (New York: Routledge, 1991). Haraway states that there are two major points in her essay: (1) that the production of totalizing theory is a major mistake that misses most of reality, and (2) "taking responsibility for the social relations of science and technology means refusing an anti-science metaphysics, a demonology of technology, and so means embracing the skillful task of reconstructing the boundaries of daily life, in partial connection with others, in communication with all of our parts" (p. 181). She suggests that cyborg imagery is a way out of the dualisms that we use to explain our bodies and technologies to ourselves. It is not based on the idea of a universal common, reductive language, but a "powerful infidel heteroglossia."

12. Kevin Kelly, ''Virtual World Sickness,'' *Whole Earth Review* (Summer 1989): 86; citing the *New York Times*, February 20, 1989.

13. Chip Morningstar and Randall Farmer, ''Cyberspace Colonies.'' Paper presented at the Second International Conference on Cyberspace, organized by the Group for the Study of Virtual Systems, Center for Cultural Studies, University of California, Santa Cruz, April 19-20, 1991. Morningstar and Farmer address the conflict of visions within the colonization and settlement of cyberspace, and discuss MUD colonies, Prodigy, Use Net, Minitel chat lines, Club Caribe, and Fujitsu Habitat as social systems within the context of centralized and decentralized networks.

14. Fred Ritchin, *In Our Own Image* (New York: Aperture, 1990), 17. *National Geographic* editor Wilbur E. Garrett referred to the modification ''not as falsification but merely the establishment of a new point of view by the retroactive repositioning of the photographer a few feet to one side.''

15. Robert Rivlin, *The Algorithmic Image* (Redmond, Wash.: Microsoft Press, 1986). Although this book does not provide a critical examination of computer imaging technology, it stands as a good introduction to the history and basic concepts of computer graphics.

16. Henri Lefebvre, *The Production of Space*, trans. Donald Nicholson-Smith (Oxford: Basil Blackwell, 1991). Lefebvre articulately examines the relationships between the abstract, mental space of the philosophers and the social and actual space of everyday life. His classic work on the mechanics of the production of space critically discusses the topic within local and global, technological, scientific, cultural, and historical projects.

17. David Harvey, *The Condition of Postmodernity* (Oxford: Basil Blackwell, 1990).

18. John Rennie, ''Living Together: Trends in Parasitology,'' *Scientific American* (January 1992): 132.

19. Ibid.

20. Adam Heilbrun, ''An Interview with Jaron Lanier,'' *Whole Earth Review* (Fall 1989): 108-19.

21. Telephone conversation with Cindy Stilwell, October 1990.

22. Morningstar and Farmer, ''Cyberspace Colonies.''

23. Carl Eugene Loeffler, ''Networked Virtual Reality and Art.'' Paper presented at the conference, ''The Simulated Presence: A Critical Response to Electronic Imaging,'' organized by the Institute for Studies in the Arts, Arizona State University, Tempe, February 19-20, 1993. Loeffler presented several research projects in which he is currently involved at Carnegie-Mellon University, where he is project director of Telecommunications and Virtual Reality and the Studio for Creative Inquiry.

Works Cited

Aristotle. *Physics*. Trans. Philip Wicksteed and Francis Cornford. London: Loeb/Heinemann, 1979.

Acker, Kathy. *Empire of the Senseless*. New York: Grove, 1988.

Augustini, Aureli. *Confessions*. Ed. P. Knoll. Leipzig: Verlag von B. G. Teubner, 1898.

Augé, Marc. "Le Fétiche et son objet," in *L'Objet en psychanalyse*. Ed. Maud Mannoni. Paris: Denöel, 1986.

Aveni, Anthony F. *Skywatchers of Ancient Mexico*. Austin: University of Texas Press, 1980.

Ballard, J. G. *Crash*. 1973; New York: Vintage, 1985.

Barlow, John Perry. "Being in Nothingness." *Mondo 2000* (Summer 1990).

Barthes, Roland. *La chambre claire: Note sur la photographie*. Paris: L'Etoile/Gallimard/Seuil, 1980.

Baudelaire, Charles. *Les paradis artificiels*. Paris: Gallimard, 1961.

Baudrillard, Jean. "The Ecstasy of Communication." In *The Anti-Aesthetic: Essays on Post-modern Culture*. Ed. Hal Foster. Port Townsend, Wash.: Bay Press, 1983.

______. *Simulacres et simulations*. Paris: Editions de Galilée, 1981.

______. *Simulations*. Trans. Paul Foss, Paul Patton, and Philip Beitchman. New York: Semiotext(e), 1983.

______. *La transparence du mal*. Paris: Galilée, 1990.

Bear, Greg. *Blood Music*. New York: Ace, 1986.

Bell, Daniel. *The Cultural Contradictions of Capitalism*. New York: Basic Books, 1978.

Benjamin, Walter. *Illuminations*. Ed. Hannah Arrendt, Trans. Harry Zohn. New York: Schocken, 1969.

______. *One Way Street and Other Writings*. Trans. E. Jephcott and K. Shorter. London: NLB, 1979.

Bersani, Leo. *The Culture of Redemption*. Cambridge, Mass.: Harvard University Press, 1990.

Blanchot, Maurice. *The Unavowable Community*. Barrytown, N.Y.: Station Hill, 1989.

Borch-Jacobsen, Mikael. *The Freudian Subject*. London: Macmillan, 1989.

Borges, Jorge Luis. "The Aleph." In *A Personal Anthology*. Ed and Trans. Anthony Kerrigan. New York: Grove, 1967.

———. *Prosa completa.* 2 vols. Barcelona: Bruguera, 1980.

Bourdieu, Pierre. *The Political Ontology of Martin Heidegger.* Trans. Peter Collier. Stanford, Calif.: Stanford University Press, 1991.

Braidotti, Rosi. "Organs without Bodies." *Differences* 1, no. 1 (1989): 147-61.

Brennan, Teresa. "Controversial Discussions and Feminist Debate." In *Freud in Exile.* Ed. Edward Timms and Naomi Segal. London: Yale University Press, 1988.

———. *History after Lacan.* London: Routledge, forthcoming.

———. *The Interpretation of the Flesh.* Vol. 1, *Freud's Theory of Feminity.* London: Routledge, 1992.

Bufo, Guiseppe. *Nicolas de Cues.* Paris: Seghers, 1964.

Bukatman, Scott. "The Cybernetic (City) State: Terminal Space Becomes Phenomenal." *Journal of the Fantastic in the Arts* 2 (1989).

———. "Who Programs You? The Science Fiction of the Spectacle." In *Alien Zone: Cultural Theory and Contemporary Science Fiction Cinema.* Ed. Annette Kuhn. London: Verso, 1990.

Ceram, C. W. *Archaeology of the Cinema.* New York: Harcourt, Brace and World, 1965.

Cioran, Emile M. *History and Utopia.* Trans. Richard Howard. New York: Seaver, 1987.

Cixous, Hélène. *Illa.* Paris: Des Femmes, 1980.

Cixous, Hélène, and Catherine Clément. *The Newly Born Woman.* Trans. Betsy Wing. Minneapolis: University of Minnesota Press, 1986.

Clarke, Bruce. "Circe's Metamorphosis: Late Classical and Early Modern Neoplatonic Readings of the *Odyssey* and Ovid's *Metamorphoses.*" *University of Hartford Studies in Literature* 21, no. 2 (1989).

Cornell, Drucilla. "Time, Deconstruction, and the Challenge to Legal Positivism: The Call for Judicial Responsibility." *Yale Journal of Law and the Humanities* 2 (Summer 1990): 67-97.

de Certeau, Michel. *L'invention du quotidien.* Paris: Bourgois, 1980.

DeHaven, Tom. *Freaks Amour.* New York: Penguin, 1986.

Deleuze, Gilles. *Le Pli.* Paris: Minuit, 1988.

———. *The Fold.* Trans. Tom Conley. Minneapolis: University of Minnesota Press, 1992.

Deleuze, Gilles, and Félix Guattari. "La géophilosophie." In *Qu'est-ce que la philosophie.* Paris: Minuit, 1991.

De Quincey, Thomas. *Confessions of an English Opium Eater.* London: Penguin, 1971.

Derrida, Jacques. "Given Time: The Time of the King." Trans. Peggy Kamuf. *Critical Inquiry* 18 (Winter 1992).

———. *Mémoires: Pour Paul de Man.* Paris: Galilée, 1989.

———. "Plato's Pharmacy." In *Dissemination.* Trans. Barbara Johnson. Chicago: University of Chicago Press, 1981.

———. "Rhétorique de la drogue." In *L'Esprit des drogues: La Dépendence hors la loi?* Ed. Jean-Michel Herviev. Paris: Autrement Revue, 1989.

———. "Two Words for Joyce." In *Post-Structuralist Joyce: Essays from the French.* Ed. Derek Attridge and Daniel Ferrer. Cambridge: Cambridge University Press, 1984.

d'Harnoncourt, Anne. *Manual of Instructions for Étant Donnés: 1 la chute d'eau, 2 le gaz d'éclairage.* Philadelphia: Philadelphia Museum of Art, 1987.

d'Harnoncourt, Anne, and Walter Hopps. *Étant Donnés: 1 la chute d'eau, 2 le gaz d'éclairage, Reflections on a New Work by Marcel Duchamp.* Philadelphia: Philadelphia Museum of Art, 1969.

Douglas, M. *Purity and Danger: An Analysis of Concepts of Pollution and Taboo.* London: Routledge and Kegan Paul, 1966.

Downs, Roger M., and David Stea. *Maps in Minds.* New York: Harper and Row, 1977.

Eagleton, Terry. "Capitalism, Modernism and Postmodernism." *New Left Review* 152 (July/August 1985).

Ehrlich, Paul. *The Population Explosion.* New York: Simon and Schuster, 1990.

Ellenberger, H. F. *The Discovery of the Unconscious: The History and Evolution of Dynamic Psychiatry.* London: Allan Lane, 1970.
Ferry, Luc, and Alain Renaut. *Heidegger and Modernity.* Trans. Franklin Philip. Chicago: University of Chicago Press, 1990.
Fisher, Scott. "Personal Simulations and Telepresence." *Multimedia Review* 1, (Summer 1990).
Flaubert, Gustave. *Madame Bovary.* Ed. Leo Bersani, Trans. Lowell Blair. Toronto: Bantam, 1981.
Friedhoff, Richard M., and William Benzon. *Visualization: The Second Computer Revolution.* New York: Harry N. Abrams, 1989.
Freud, Sigmund. *Beyond the Pleasure Principle.* In *The Standard Edition of the Complete Psychological Works*, vol. 18. London: Hogarth Press and the Institute of Psychoanalysis, 1953-74. 24 vols.
______. *Inhibitions, Symptoms and Anxiety.* In *The Standard Edition of the Complete Psychological Works*, vol. 20.
______. *The Interpretation of Dreams.* In *The Standard Edition of the Complete Psychological Works*, vol. 5.
______. *Project for a Scientific Psychology (Entwurf einer Psychologie).* In *The Standard Edition of the Complete Psychological Works*, vol. 1.
______. *Studies on Hysteria.* In *The Standard Edition of the Complete Psychological Works*, vol. 2.
Gadamer, Hans-Georg. "Back to Syracuse." *Critical Inquiry*, 15 (Winter 1989): 427-31.
______. "Rainer Maria Rilke Deutung des Daseins." In *Kleine Schriften*, vol. 2. Tübingen: J. C. B. Mohr.
Gautier, Théophile. *La Pipe d'opium, Le Hachich, Le Club des Hachichins.* Paris: Gallimard, 1961.
Gibson, William. *Neuromancer.* New York: Ace, 1984.
Gibson, William, and Bruce Sterling. *The Difference Engine.* New York: Bantam Spectra, 1991.
Giddens, Anthony. *The Consequences of Modernity.* Stanford, Calif.: Stanford University Press, 1990.
Giedion, Siegfried. "Mechanization and Death: Meat." In *Mechanization Takes Command: A Contribution to Anonymous History.* New York: Norton, 1969.
Gilligan, Carol. *In a Different Voice: Psychological Theory and Women's Development.* Cambridge, Mass.: Harvard University Press, 1982.
Glissant, Edouard. *Le discours antillais.* Paris: Seuil, 1980.
Glover, E. "On the Aetiology of Drug-Addiction." *International Journal of Psychoanalysis* 13 (1932).
Greenland, Colin. "A Nod to the Apocalypse: An Interview with William Gibson." *Foundation* 36 (Summer 1986).
Grosz, Elizabeth. "Space, Time and Bodies." In *On the Beach.* Sydney: 1988.
Guattari, Félix. *Les trois écologies.* Paris: Galilée, 1990.
Habermas, Jürgen. *Der philosophische Diskurs der Moderne: Zwölf Vorlesungen.* Frankfurt am Main: Suhrkamp Verlag, 1985.
______. *Nachmetaphysisches Denken: Philosophische Aufsätze.* Frankfurt am Main: Suhrkamp Verlag, 1988.
______. *The Philosophical Discourse of Modernity: Twelve Lectures.* Trans. Frederick Lawrence. Oxford: Polity, 1987.
Hafner, Katie, and John Markoff. *Cyberpunk: Outlaws and Hackers on the Computer Frontier.* New York: Simon and Schuster, 1991.
Hamilton, Richard. "The Green Book" (London, 1960). In *The Bride Stripped Bare by Her Bachelors, Even.* Richard Hamilton and George Heard Hamilton. New York: Jaap Rietman, 1976.
Haraway, Donna. "Cyborg Manifesto." In *Simians, Cyborgs, and Women: The Reinvention of Nature.* New York: Routledge, 1991.

———. "A Manifesto for Cyborgs: Science, Technology, and Socialist Feminism in the 1980's." In *Feminism/Postmodernism*. Ed. Linda J. Nicholson. New York: Routledge, 1990.

———. "A Manifesto for Cyborgs: Science, Technology, and Socialist Feminism in the 1980's." *Socialist Review* 80 (1985): 65-107.

Hardison, O. B., Jr. *Disappearing through the Skylight: Culture and Technology in the Twentieth Century*. New York: Viking, 1989.

Harries, Karsten. "Heidegger as a Political Thinker." In *Heidegger and Modern Philosophy: Critical Essays*. Ed. Michael Murray. New Haven, Conn.: Yale University Press, 1978.

Harvey, David. *The Condition of Postmodernity*. Oxford: Basil Blackwell, 1989.

Hayles, N. Katherine. "Postmodern Parataxis: Embodied Texts, Weightless Information." *American Literary History* 2, no. 3 (1990).

Heidegger, Martin. *Basic Writings*. Ed. David Farrell Krell. New York: Harper and Row, 1977.

———. *Being and Time*. Trans. John Macquarrie and Edward Robinson. Oxford: Basil Blackwell, 1962.

———. "Conversation on a Country Path about Thinking." In *Discourse on Thinking*. Trans. John Anderson and E. Hans Freund. New York: Harper and Row, 1966.

———. "Das Seins des Daseins als Sorge." In *Sein und Zeit*. Tübingen: Max Niemeyer Verlag, 1986.

———. *Discourse on Thinking: A Translation of Gelassenheit*. Trans. John Anderson and E. Hans Freund. New York: Harper and Row, 1966.

———. *Essais et conférences*, vol. 1. Paris: Gallimard, 1988.

———. "Letter on Humanism." In *Basic Writings*. Ed. David Farrell Krell. New York: Harper and Row, 1977.

———. "Logos" (Heraclitus, Fragment B 50). In *Early Greek Thinking*. Trans. David Farrell Krell and Frank A. Capuzzi. New York: Harper and Row, 1975.

———. *The Principle of Reason*. Trans. Reginald Lilly. Bloomington: Indiana University Press, 1991.

———. *The Question Concerning Technology and Other Essays*. Trans. W. Lovitt. New York: Harper and Row, 1977.

———. "The Self-Affirmation of the German University." *Review of Metaphysics* 38 (1985).

———. "What Are Poets For?" In *Poetry, Language, Thought*. Trans. Albert Hofstadter. New York: Harper and Row, 1971.

———. *What Is a Thing?* Trans. W. B. Barton and Vera Deutsch. South Bend, Ind.: Regnery/Gateway, 1967.

———. "Zur Erörterung der Gelassenheit: Aus cinem Feldweggespräch über das Denken." In *Gelassenheit*. Pfüllingen: Verlag Günther Neske, 1959.

Heilbrun, Adam. "An Interview with Jaron Lanier." *Whole Earth Review* (Fall 1989): 108-19.

Henderson, Joseph. "Designing Realities: Interactive Media, Virtual Realities and Cyberspace." *Multimedia Review* 1 (Summer 1990).

Hold, Klaus. "The Finitude of the World: Phenomenology in Transition from Husserl to Heidegger." Trans. Anthony Steinbock. Paper presented at the Goethe Institute conference, "Heidegger: Philosophy, Art, and Politics," sponsored by the British Society for Phenomenology, January 1990.

Irigaray, Luce. *Ethique de la différence sexuelle*. Paris: Minuit, 1984.

———. *Speculum of the Other Woman*. Trans. Gillian C. Gill. Ithaca, N.Y.: Cornell University Press, 1985.

———. *This Sex Which Is Not One*. Trans. Catherine Porter and Carolyn Burke. Ithaca, N.Y.: Cornell University Press, 1985.

Jameson, Fredric. *The Political Unconscious*. Ithaca, N.Y.: Cornell University Press, 1981.

———. "Postmodernism, or The Cultural Logic of Late Capitalism." *New Left Review* 146 (July/August 1984): 59-92.

______. *Postmodernism, or The Cultural Logic of Late Capitalism*. Durham, N.C.: Duke University Press, 1991.

Jay, M. "In the Empire of the Gaze: Foucault and the Denigration of Vision in Twentieth-Century French Thought." In *Foucault: A Critical Reader*. Ed. D. Couzens Hoy. Oxford: Basil Blackwell, 1986.

Joyce, James. *Finnegans Wake*. Harmondsworth: Penguin, 1986.

Jünger, Ernst. *Annäherungen: Drogen und Rausch*. Stuttgart: Ernst Klett Verlag, 1970.

Keller, Evelyn Fox. "Feminism and Science." *Signs* (Spring 1982): 589-607.

Kelly, Kevin. "Virtual Reality: An Interview with Jaron Lanier." *Whole Earth Review* 64 (Fall 1989).

______. "Virtual World Sickness." *Whole Earth Review* (Summer 1989).

Klée, Paul. *Théorie de l'art moderne*. Paris: Gonthier, 1963.

Klein, Melanie. "Early Stages of the Oedipus Conflict." In *Love, Guilt and Reparation and Other Works, 1921-1945: The Writings of Melanie Klein*, Vol. 1. London: Hogarth, 1985.

______. "Envy and Gratitude." In *Envy and Gratitude and Other Works, 1946-1963: The Writings of Melanie Klein*, Vol. 3. London: Hogarth, 1980.

______. "Notes on Some Schizoid Mechanisms." In *Love, Guilt and Reparation and Other Works, 1921-1945: The Writings of Melanie Klein*, Vol. 1. London: Hogarth, 1985.

Kristeva, J. "L'abjet d'amour." *Tel Quel* 91 (1981): 17-32.

Krueger, Myron W. *Artificial Reality*. Reading, Mass.: Addision-Wesley, 1983.

______. "Artificial Reality: Past and Future." *Multimedia Review* 1 (Summer 1990).

LaCapra, Dominick. *"Madame Bovary" on Trial*. Ithaca, N.Y.: Cornell University Press, 1982.

Lacan, Jacques. "Aggressivity in Psychoanalysis." In *Ecrits: A Selection*. Trans. Alan Sheridan. New York: Norton, 1977.

______. *Ecrits: A Selection*. Trans. Alan Sheridan. New York: Norton, 1977.

______. *Encore: Le séminaire XX*. Ed. Jacques-Alain Miller. Paris: Seuil, 1975.

______. "The Function and Field of Speech and Language in Psychoanalysis." In *Ecrits: A Selection*. Trans. Alan Sheridan. New York: Norton, 1977.

______. "Some Reflections on the Ego." *International Journal of Psychoanalysis* 34, no. 1 (1953).

Laplanche, J. *Life and Death in Psychoanalysis*. Trans. J. Mehlman. Baltimore: Johns Hopkins University Press, 1976.

Laplanche, J., and J.-P. Pontalis. "Fantasy and the Origins of Sexuality." *International Journal of Psychoanalysis* 49 (1968): 1-18.

Lasch, Christopher. *The Culture of Narcissism*. New York: Norton, 1978.

Lefbvre, Henri. *The Production of Space*. Trans. Donald Nicholson-Smith. Oxford: Basil Blackwell, 1991.

______. "Le Fétiche et son objet." In *L'Objet en psychanalyse*. Ed. Maud Mannoni. Paris: Denoël, 1986.

Le Poulichet, Sylvie. *Toxicomanies et Psychanalyse: Les Narcoses du Désir*. Paris: Presses Universitaires de France, 1987.

Levinas, Emmanuel. *Sur Maurice Blanchot*. Paris: Fata Morgana, 1975.

Lipovetsky, Gilles. *L'Empire de l'éphémère: la mode et son destin dans les sociétés modernes*. Paris: Gallimard, 1987.

Loeffler, Carl Eugene. "Networked Virtual Reality and Art." Paper presented at the conference, "The Simulated Presence: A Critical Response to Electronic Imaging," organized by the Institute for Studies in the Arts, Arizona State University, Tempe, February 19-20, 1993.

Luhmann, Niklas. *The Differentiation of Society*. New York: Columbia University Press, 1982.

MacCannell, Juliet. *The Regime of the Brothers*. London: Routledge, 1991.

Morningstar, Chip, and Randall Farmer. "Cyberspace Colonies." Paper presented at the Second International Conference on Cyberspace, Santa Cruz, Calif., April 19-20, 1991.

Naess, Arne. *Ecology, Community, and Lifestyle: Outline of an Ecosophy*. Trans. David Rothenberg. Cambridge: Cambridge University Press, 1989.

Nancy, Jean-Luc. *The Inoperative Community*. Ed. Peter Connor. Minneapolis: University of Minnesota Press, 1991.

Nichols, Bill. "The Work of Culture in the Age of Cybernetic Systems." *Screen* 29 (Winter 1988).

Nietzsche, Friedrich. *The Gay Science*. New York: Penguin, 1986.

Paz, Octavio. *El arco y la lira*, 3d ed. Mexico: Fondo de Cultura Económica, 1986.

———. "Poetry and the Free Market," *New York Times*, November 1991.

Penrose, Roger. *The Emperor's New Mind: Concerning Computers, Minds, and the Laws of Physics*. New York: Oxford University Press, 1989.

Prigogine, Ilya, and Isabelle Stengers. *La nouvelle alliance*. Paris: Gallimard, 1979.

Prigogine, Ilya, and Isabelle Stengers. *Order out of Chaos* (foreword by Alvin Toffler). Boulder, Colo.: New Science Library, 1984.

Regis, Ed. *Great Mambo Chicken and the Transhuman Condition: Science Slightly over the Edge*. Reading, Mass.: Addison-Wesley, 1990.

Rennie, John. "Living Together: Trends in Parasitology." *Scientific American* (January 1992).

Rheingold, Howard. "Teledildonics: Reach Out and Touch Someone." *Mondo 2000* (Summer 1990).

———. *Virtual Reality: The Revolutionary Technology of Computer-Generated Artificial Worlds and How It Promises and Threatens to Transform Business and Society*. New York: Simon and Schuster, 1991.

Ricoeur, Paul. "Metaphor and the Central Problem of Hermeneutics." In *Hermeneutics and the Human Sciences: Essays on Language, Action and Interpretation*. Ed. and Trans. Jane B. Thompson. Cambridge: Cambridge University Press, 1981.

Rilke, Rainer Maria. *Duino Elegies*. Trans. J. B. Leishman and Stephen Spender. New York: Norton, 1963.

Ritchin, Fred. *In Our Own Image*. New York: Aperture, 1990.

Rivlin, Robert. *The Algorithmic Image*. Redmond, Wash.: Microsoft Press, 1986.

Ronell, Avital. *Crack Wars: Literature/Addiction/Mania*. Lincoln: University of Nebraska Press, 1991.

Rose, Jacqueline. "Introduction II." In *Feminine Sexuality: Jacques Lacan and the Ecole Freudienne*. Ed. Juliet Mitchell and Jacqueline Rose. London: Macmillan, 1982.

Schürmann, Reiner. *Heidegger, on Being and Acting: From Principles to Anarchy*. Trans. Christine-Marie Gros. Bloomington: Indiana University Press, 1987.

———. "Principles Precautious: On the Origin of the Political in Heidegger." In *Heidegger, the Man and the Thinker*. Ed. Thomas Shennhan. Chicago: Precedent, 1981.

Serres, Michel. *Hermes: Literature, Science, Philosophy*. Ed. Josué V. Harari and David F. Bell. Baltimore: Johns Hopkins University Press, 1982.

———. *Le contrat naturel*. Paris: Bourin, 1990.

Shelley, Mary. *Frankenstein*. New York: Bantam, 1981.

Shiva, Vandana. *Staying Alive*. London: AEC, 1989.

Sloterdijk, Peter. "Dionysus Meets Diogenes; or, The Adventures of the Embodied Intellect." In *Thinker on Stage: Nietzsche's Materialism*. Trans. Jamie Owen Daniel. Minneapolis: University of Minneosta Press, 1989.

Sobchack, Vivian. "The Scene of the Screen: Toward a Phenomenology of Cinematic and Electronic 'Presence.' " Ed. Hans U. Gumbrecht and K. Ludwig Pfeiffer. In *Materialität der Kommunikation*. Frankfurt am Main: Suhrkamp, 1988.

———. *Screening Space: The American Science Fiction Film*, 2d ed. New York: Ungar, 1988.

Spring, Michael. "Informating with Virtual Reality." *Multimedia Review* 1 (Summer 1990).

Stanworth, Michelle, ed. *Reproductive Technologies*. Cambridge: Polity, 1987.

Steegmuller, Francis. *Flaubert and Madame Bovary: A Double Portrait*. Chicago: University of Chicago Press, 1977.

Stone, Allucquere Rosanne. "Will the Real Body Please Stand Up? Boundary Stories about Virtual Cultures." In *Cyberspace: First Steps*. Ed. Michael Benedikt. Cambridge: MIT Press, 1991.

Tannoudji, G. Cohen, and M. Spiro. *La matière-espace-temps*. Paris: Fayard, 1986.

Trusted, Jennifer. *Physics and Metaphysics*. London: Routledge, 1991.

Varela, Francisco. *Autonomie et connaissance*. Paris: Seuil, 1989.

Vinge, Vernon. *True Names and Other Dangers*. New York: Baen, 1987.

Virilio, Paul. *Défense populaire et luttes écologiques*. Paris: Galilée, 1978.

______. *L'espace critique*. Paris: Bourgois, 1984.

______. *L'inertie polaire*. Paris: Christian Bourgois, 1990.

Virilio, Paul, and Sylvère Lotringer. *Pure War*. Trans. Mark Polizzotti. New York: Semiotext(e), 1983.

Walker, John. "Through the Looking Glass: Beyond 'User' Interfaces." *CADalyst* (December 1989).

Walser, Randall. "On the Road to Cyberia: A Few Thoughts on Autodesk's Initiative." *CADalyst* (December 1989).

Wiener, Norbert. *Cybernetics; or, Control and Communication in the Animal and the Machine*. Cambridge, Mass.: Technology Press, 1948.

Wilden, Anthony. *The Rules Are No Game: The Strategy of Communication*. London: Routledge and Kegan Paul, 1987.

Wyschogrod, Edith. *Spirit in Ashes: Hegel, Heidegger, and Man-Made Mass Death*. New Haven, Conn.: Yale University Press, 1985.

Zimmerman, Michael. *Heidegger's Confrontation with Modernity: Technology, Politics, Art*. Bloomington: Indiana University Press, 1990.

Contributors

Teresa Brennan is lecturer in the Faculty of Social and Political Sciences at Cambridge University. She is the editor of *Between Feminism and Psychoanalysis* and author of *The Interpretation of the Flesh* (Routledge, 1992).

Patrick Clancy is a video artist and chair of the Video Department at Kansas City Art Institute. His one-man show and performance *365/360* is currently touring the country. He is also a mycologist of great repute.

Scott Durham is assistant professor of French at Northwestern University.

Jennifer Gage is a graduate student in French at Brown University.

Françoise Gaillard is professor of social sciences University of Paris VII. She guest teaches at Brown University and is known for her writings on literature and science in nineteenth-century France.

Félix Guattari, whose writings on technology and ecology inspired much of this volume, dedicated his life to fighting institutional powers at all levels. In recent years, he emphasized the importance of ecological thinking and tried to enter French politics as a Green. He believed in thinking transversally and in bringing together science and the humanities. One of his latest books, *Chaosmose* (Galilée, 1991), develops further the theses put forth in his contribution to this volume.

N. Katherine Hayles is Carpenter Professor of English at the University of California, Los Angeles. She writes on literature and science in the twentieth century, and is currently at work on a book titled *Virtual Bodies: Literature in the Age of Information*.

Jeffrey S. Librett teaches in the Department of Modern Languages and Literatures at Loyola University of Chicago. He has published articles on de Man, F. Schlegel, Schiller, and Kant, and has translated several critical essays from French and German, including the collection *Of the Sublime: Presence in Question* (SUNY Press), by Jean-Luc Nancy et al.

Alberto Moreiras teaches Latin American literature at Duke University. He has published *Interpretación y diferencia* (Madrid) and a number of essays on contemporary Hispanic American fiction and literary theory.

Jean-Luc Nancy participated in the first L. P. Irvin Colloquium on community, the topic of which was inspired by his book *The Inoperative Community* (University of Minnesota Press). He wrote the essay that appears in this volume from his hospital bed, while recovering from heart transplant surgery.

Avital Ronell is professor of comparative literature at the University of California, Berkeley. Her recent publications include *The Telephone Book* and *Crack Wars* (both University of Nebraska Press).

Ingrid Scheibler is a research fellow at Newnham College, Cambridge. She is the author of a book on Heidegger and feminist theory titled *Earth's Destiny* (Routledge, forthcoming).

Paul Virilio, trained as an urban architect, is a leading culture critic. He sometimes lectures at the Collège International de Philosophie. He is fascinated by technology and dismayed by the fact that it has not been able to transform social space in adequate ways. His numerous books include *Bunker Archéologie, Logistique de la perception, Politique de la Vitesse, L'horizon négatif*, and *Défense populaire et luttes écologiques* (all Galilée). Translations have appeared in North America in a special series edited by Sylvère Lotringer.

The Miami Theory Collective is based in the Department of French and Italian at Miami University in Oxford, Ohio, but includes members from other areas of knowledge. The Collective sponsors sustained and focused dialogue on issues it considers important to contemporary discourse. The Miami Theory Group for this volume consisted of Verena Andermatt Conley, Tom Conley, James Creech,

Steve Nimis, Linda Singer, and Marie-Claire Vallois. George Van den Abbeele was also present in the beginning stages of *Rethinking Technologies*.

Tom Conley is professor of French, University of Minnesota, and visiting professor, Miami University. He is currently working on the relation between mapping and early modern writing.

Verena Andermatt Conley is professor of French and Women's Studies at Miami University. She is at work on problems of feminism, ecology, and technology in contemporary culture.

James Creech is professor of French at Miami University. He is interested inissues in gay studies.

Stephen Nimis is professor of classics at Miami University. He is also interested in contemporary issues of technology.

Linda Singer was professor of philosophy at Miami University before her untimely death. Her book *Erotic Welfare* (Routledge), edited by Judith Butler, is forthcoming.

Marie-Claire Vallois is professor of French at Miami University. She writes on issues concerning women and the Revolution. She has published *Fictions féminines: Mme de Staël, les voix de la Sybille*.

Index

Acker, Kathy, 187
addiction, 59-66, 69
Adorno, Theodor, 144
allopoiesis, 16
Aristotle, 13, 41, 106
atopia, 4, 86
Augé, Marc, 21
Augustine, Saint, 99, 100
auratic object, 160, 161
autopoiesis, 13, 15, 17, 24
autopoietic nexus, 14-16

Ballard, J. G., 158, 159, 162-64, 167
Barlow, John Perry, 180n
Barthes, Roland, 195
Baudelaire, Charles, 62
Baudrillard, Jean, 83-86, 89, 159-67, 174
becoming, 22, 81, 84, 85, 89, 219
Being and Time (Heidegger), 59, 93, 118, 121, 122, 124
Bell, Daniel, 143-44, 152, 154
Benjamin, Walter, 88, 92, 93, 110, 144, 158, 160
Bersani, Leo, 113*n*38
biosphere, 17, 25
Blanchot, Maurice, 61
Borch-Jacobsen, Mikael, 112*n*16
Borges, Jorge Luis, 194-97, 199, 200
Braidotti, Rosi, 112*n*15
Bukatman, Scott, 186
Burroughs, William, 186

capitalism, xi, 80, 81, 86, 143, 147, 175
Cartesian: grid, 212; terms, 219
Celan, Paul, 80
Certeau, Michel de, 91*n*23
chaos theory, 84
Cioran, Emile, 147
Cixous, Hélène, 78-80, 82, 88, 89, 187
Clausewitz, Baron von, 40
commodity, 93-99, 101, 103, 107, 110, 113*n*35
community, xiv*n*5, 43, 44, 48, 51, 168
Conley, Tom, 28
consensus, xi, 53, 145
consumer, 93, 153
consumerism, ix, xi
Cornell, Drucilla, 113*n*37
Crash (Ballard), 158, 162-65
Cronenberg, David, 186
cybernetics, 174, 176, 188, 192, 195

cyberspace, x, xiv, 176, 177, 185, 187, 195, 197-99, 202, 207, 218, 221, 222, 225, 230

Dasein, 59, 60, 63, 122
Debray, Régis, 145
DeHaven, Tom, 182
democracy, 37, 44, 153, 154
De Quincey, Thomas, 65
Derrida, Jacques, 73*n*13, 79, 113*n*37, 119, 198, 203*n*13, 204*nn*29, 35
Descartes, René, 14
deterritorialization, 14, 16, 25
désoeuvrement, 61
diagram, 18, 20
diagrammatic: feedback, 26; semiotic components, 14; semiotic machine, 15; semiotization, 20-22; virtualities, 18
Dick, Philip K., 61, 183
Ding-an-sich/Ding-an-mich, 139*n*70
Douglas, Mary, 112*n*22
drugs, ix, 61, 64-67
Duchamp, Marcel, 144, 214, 215
Duras, Marguerite, 61

Eagleton, Terry, 169-70*n* 11
East-West axis, 35
ecology, xi, xiii, 52, 77, 80, 82, 83, 85, 86, 216
ecotechnics, 51-53, 55
ego, 25, 94, 95, 99-102, 105-7, 110
Eisenstein, Sergei, 212
enframing, xi, xii, 115, 125
environment, ix, x, xiii, 4, 8, 10, 11, 83, 86, 93, 109, 110, 128, 177, 207, 210, 215, 216, 226
ethos, 32
exposure, 7

Farmer, Randall, 204*n*13
Fisher, Scott, 8
Flaubert, Gustave, 64, 66
Foucault, Michel, 114*n*40, 119
foundation, 31, 33, 36, 49, 54, 101, 129
foundational psychical fantasy, 99-102, 109-10, 111*n*3
Fraisse, Robert, 29
Frampton, Hollis, 213, 220
Freud, Sigmund, 48, 64, 94-97, 99, 103-5, 107

Gadamer, Hans-Georg, 118
genocide, xiii
geocide, xiii
geopolitics, 9
Gestell, xii, 59, 60, 63, 125
Gibson, William, 176, 178, 192
Giddens, Anthony, 113*n*37
Giedion, Siegfried, 156-58
gift, 198, 200, 201
Glissant, Edouard, xiv*n*2, 90*n*3
Godard, Jean-Luc, 77, 78
Griffith, D. W., 213
Grosz, Elizabeth, 114*n*
Guattari, Félix, xiii, xiv*n*3, 87, 88
Gulf War, 33, 34, 43, 45, 60, 78, 226, 227, 229

Habermas, Jürgen, 117-22, 124-26, 131-33
habitus, 32
Haraway, Donna, 111-12*n*15, 175, 193, 231*n*11
Hardison, O. B., Jr., 188*n*1
Harnoncourt, Anne d', 231*n*9
Harvey, David, 81-82, 86, 88, 89, 232*n*17
Hegel, Georg Wilhelm Friedrich, 64, 100, 154
Hegelian: dialectics, 78
Heidegger, Martin, x, xii, xiii, 22, 62, 70, 78, 80, 93, 100, 115-39, 203*n*2
Heideggerian: mode of philosophy 13; ontology, 26
heterogenesis, 16, 24
history, ix, 17, 34, 47, 48, 50, 54, 61, 222, 225
Husserl, Edmund, 120

ideology, x, xii
inertia, 11, 105, 106
intensity, 22, 25, 26
interface, 4, 10
interval, 6, 10
Irigaray, Luce, 78, 114*n*41, 187

Jameson, Fredric, 169*n*11, 174, 213, 214
Jay, Martin, 114*n*40
Joyce, James, 204*n*40

Kant, Immanuel, 65, 70, 152, 213
Keller, Evelyn Fox, 114*n*41
Kelly, Kevin, 231*n*12
"kitschization," 147, 151

Klee, Paul, 4, 8, 168-69*n*2
Klein, Melanie, 94-96, 99-101
Klein, Yves, 220
Kristeva, Jùlia, 101
Krueger, Myron W., 189*n*8

Lacan, Jacques, 94, 96, 99-101, 105, 106, 109, 186, 187
Lacanian signifier, 16, 20, 23
LaCapra, Dominick, 72-73*n*10
Lanier, Jaron, 183, 227
Laplanche, Jean, 104, 105
Lasch, Christopher, 143
Lautréamont, Isidore Ducasse, 212
Lefebvre, Henri, 232*n*16
LeGuin, Ursula, 180
Leibniz, Gottfried Wilhelm, 27*n*4, 192
Leroi-Gourhan, André, 14
Lévi-Strauss, Claude, 91*n*26
Lipovetsky, Gilles, 147-48, 154
Loeffler, Carl Eugene, 227, 232*n*23
Luhman, Niklas, 113*n*37

McCannell, Juliet, 112*n*23
machine, 13-18, 20, 22, 25-27, 61, 208, 209
machinic ordering, 14, 18, 24-27
machinism, 13, 15, 17, 18
Madame Bovary (Flaubert), 66
Malthus, Thomas, xii
Marx, Karl, 35, 113*n*35
mecanosphere, 17, 25
media, ix, xiii, 61
Merleau-Ponty, Maurice, 114*n*40
mimesis, 15, 41
Mondo 2000, 181
Morningstar, Chip, 232*n*13

Nancy, Jean-Luc, 61
National Geographic, 223
nationalism, 31, 53
nature, xi, xiii, 13, 40-42, 79, 80, 98-101, 109, 207, 208, 217, 225
neo-Kantian humanism, 34
Neuromancer (Gibson), 176, 182, 185
Newton, Isaac, 6
Newtonian immobility and timelessness, 80
Nicholas of Cusa, 9
Nichols, Bill, 175, 183, 188
Nietzsche, Friedrich, 62, 63, 65, 70, 72, 118, 119

North-South axis, xii, 30, 35

ontogenesis, 17
ontogenetic thresholds, 24

Paz, Octavio, 151, 191
Peirce, Charles Sanders, 20
Penrose, Roger, 188*n*1
philogenetic thresholds, 24
phusis, 40-43, 110
phylum, 15, 17, 21, 23
Pickover, Clifford, 223
poiesis, xii, xiii, 78, 79, 115
Pontalis, Jean-Baptiste, 104
present time, 4
Prigogine, Ilya, xi, xiv*n*4, 84, 89
Pynchon, Thomas, 113*n*38, 184, 185

Ray, Thomas, 225
relativity, 3, 5, 6, 25
Renaissance, x
Rheingold, Howard, 181, 197
Ricoeur, Paul, 137*n*37
Rilke, Rainer Maria, 123, 137*n*33
Rousseau, Jean-Jacques, 32

Shelley, Mary, 208, 209
Shiva, Vandana, 81, 82
signifier, 16, 23, 29, 80, 180, 186, 234
simulacrum, 71, 160, 161, 163, 165, 174, 179, 219
simulation, 162, 168
simulation model, 158, 160, 161-63, 167, 168, 174, 182
Sobchack, Vivian, 189-90*n*26
sovereignty, xiii, 29, 32-38, 40, 43-58
space, 3, 5-8, 11, 12, 16, 24, 180, 201, 213, 217-19, 222
Spinoza, Benedict, 41
Stanworth, Michelle, 111*n*15
Stengers, Isabelle, xi, xiv*n*4, 84, 89
Sterling, Bruce, 192
Stilwell, Cindy, 227
Stone, Allucquere Rosanne, 202
subjectivity, xi, xiii, 14, 20, 21, 132, 152, 178, 179, 221

techné, xii, xiii, 13, 16, 26, 29, 40, 41, 43-46, 48, 49, 52, 56-58, 78, 79, 115
technology, ix, x, xi, xii, xiv, 3, 4, 9, 17, 28, 38-43, 46, 52, 56-61, 67, 80, 86, 90, 94,

102, 103, 115, 125, 131, 156, 168, 174-76, 178, 187, 191-93, 198, 199, 209, 210, 217, 218, 225, 226
telepresence, 4, 7, 8
teletechnolgies, 3, 7, 12
teletopia, 4
teletopical commutation, 6
Thom, René, 24
Thomas Aquinas, 44
time, 3, 6, 7, 11, 12, 18, 24, 25
Tinguely, Jean, 18

utopia, 4, 88

Varela, Francisco, 13, 16, 17, 18
Virilio, Paul, x, xii, xiii, 73*n*16, 78, 80, 83, 86, 87, 184, 185
virtual reality, ix, xii, 174, 175, 177, 180, 182, 185, 191, 192, 194, 203, 218, 221, 228

war, 28-39, 43-49, 52, 56, 179
Wiener, Norbert, 13, 174
Wilden, Anthony, 190
Wittgenstein, Ludwig, 184

Zahn, Johannes, 213